# Strive for a 5:
# Preparing for the AP* Statistics Examination

to accompany

**The Practice of Statistics, Fourth Edition**

Starnes, Yates, Moore

**Jason M. Molesky**

*Lakeville Area Public Schools*

ISBN-13: 978-1-4292-6239-2
ISBN-10: 1-4292-6239-7



Printed in the United States of America

First Printing

W.H. Freeman and Company
41 Madison Avenue
New York, NY 10010
Houndmills, Basingstoke RG21 6XS, England

www.whfreeman.com

## TABLE OF CONTENTS

# Preface

***Strive for a 5: Preparing for the AP* Statistics Examination*** is designed for use with *The Practice of Statistics for AP*, Fourth Edition* by Daren Starnes, Dan Yates, and David Moore. It is intended to help you evaluate your understanding of the material covered in the textbook, reinforce the key concepts you need to learn, develop your conceptual understanding and communication skills, and prepare you to take the AP Statistics Exam. This book is divided into two sections: a study guide and a test preparation section.

## THE STUDY GUIDE SECTION

The study guide is designed for you to use throughout your course in AP Statistics. As each chapter is covered in your class, you can use the study guide to help you identify and learn the important statistics concepts and terms. For each section, this guide provides learning targets, checks for understanding, practice problems, and hints to help you master the material and verify your understanding before moving on to the next section.

This guide is organized to follow the formative assessment process. It provides a means for you to clearly identify what it is you need to learn, whether or not you have learned it, and what you need to do to close any gaps in your understanding. That is, it will help you answer the questions:

"Where am I going?".
"Where am I now?". and
"How can I close the gap?"

For each chapter, the study guide is organized as follows.

- **Overview**: The chapter content is summarized to provide a quick overview of the material covered. The material is broken down into the specific learning targets to help you answer the question "Where am I going?"
- **Section content:** Each section includes a presentation of the important content to help guide your learning. The section content includes the following.
    - **Checks for Understanding**: Mini formative assessments for each learning target help you answer the question "Where am I now?"
    - **Complete the Exercise**: This exercise is designed to illustrate how to apply the statistical content developed in the section.
    - **Section Review**: A review of the key concepts from the section is provided.
- **Chapter Review**: The key concepts of the chapter are briefly reviewed.
- **Multiple Choice Questions**: These multiple-choice questions focus on the key concepts that you should grasp after reading the chapter=and are designed for quick exam preparation.
- **FRAPPY!:** A "Free-Response AP Problem, Yay!" is provided after each chapter. Each FRAPPY is modeled after actual AP Statistics Exam questions and provide an opportunity for you to practice your communication skills to maximize your score. These problems are meant to help you determine what further work is needed to address the question "How can I close the gap?"
- **Vocabulary Crossword Puzzle**: These puzzles provide a fun way for you to check your understanding of key definitions from the chapter.
- **Answer Key:** Answers to all questions and problems in the study guide are available in the back of the book. Checking your answers will help you determine whether or not you need additional work on specific learning targets.

## THE TEST PREPARATION SECTION

The "Preparing for the AP Statistics Examination" section of this guide is meant to help you better understand how the exam is constructed and scored, how best to study for the exam, and how to make sure you highlight what you have learned when answering exam questions. Two practice exams and answer keys are included to help you get a feel for the actual test.

I hope that your use of "Strive for a 5: Preparing for the AP* Statistics Examination" will assist you in your study of statistical concepts, help you earn as high a score as possible on the exam, and provide you with an interest and desire to further your study of statistics. I'd like to thank Daren Starnes and Ann Heath for inviting me to create this guide, Steven Miller and James Bush for their thoughtful reviews, and Dora Figueiredo for her guidance throughout the process.

Most importantly, I'd like to thank my wife, Anne, for her patience and support through all of my stats projects, my kiddos, Addison and Aidan, for providing much needed play breaks, and my trusty beagle, Snickers, for his company as I typed away in my office over the past few months. Best wishes for a great year of studies and best of luck on your AP Statistics Exam!

**Jason Molesky**
"StatsMonkey"

## About the Author

**Jason M. Molesky**
Assessment & Accountability Coordinator
Lakeville Area Public Schools

Jason has served as an AP Statistics Reader since 2007. He has taught AP Statistics since 2002 and is currently the Assessment and Accountability Coordinator for Lakeville schools, overseeing his district's testing program and providing professional development in formative assessment strategies, data-informed instruction, grading practices, and mathematics education. He also creates statistics resources for a variety of high school and college textbooks, teaches Assessment and Evaluation for St. Mary's University, and maintains the "StatsMonkey" website, a clearinghouse for AP Statistics resources. In his spare time, Jason enjoys spending time with his wife, their two children, and his trusty beagle.

# Chapter 1: Exploring Data

*"Statistical thinking will one day be as necessary for efficient citizenship as the ability to read and write." H.G. Wells*

## Chapter Overview

Statistics is the science of data. We begin our study of the subject by mastering the art of examining and describing sets of data. This chapter introduces you to the concept of exploratory data analysis. You will learn how to use a variety of graphical tools to display data as well as how to describe data numerically.

By the end of this chapter, you should understand the difference between categorical and quantitative data, how to display and describe these two types of data, and how to move from data analysis to inference. These skills are the basis for our entire study of Statistics. Use this guide to help ensure you have a firm grasp on these concepts!

## Sections in this Chapter



## Plan Your Learning

Use the following *suggested* guide to help plan your reading and assignments.
Note: your teacher may schedule a different pacing. Be sure to follow his or her instructions!

| **Read** | Overview xx-xxii<br>Intro: pp 2-6 | 1.1: pp 8-12 | 1.1: pp 12-22 | 1.2: pp 27-34 |
|---|---|---|---|---|
| **Do** | 1, 3, 5, 7, 8 | 11,13,15,17 | 19, 21, 23, 25, 27-32 | 37, 39, 41, 43, 45, 47 |

| **Read** | 1.2: pp 35-42 | 1.3: pp 50-58 | 1.3: pp 58-69 | Chapter Summary |
|---|---|---|---|---|
| **Do** | 53, 55, 57, 59, 60, 69-74 | 79, 81, 83, 87, 89 | 91, 93, 95, 97, 103, 105, 107-110 | Multiple Choice FRAPPY! |

## Introduction - Data Analysis: Making Sense of Data

### Before You Read: Section Summary

In this section, you will learn the basic terms and definitions necessary for our study of statistics. You will also be introduced to the idea of moving from data analysis to inference. One of our goals in statistics is to use data from a representative sample to make an inference about the population from which it was drawn. In order to do this, we must be able to identify the type of data we are dealing with as our choice of statistical procedures will depend upon that distinction.

### "Where Am I Going?"

**Learning Targets:**

____ I can identify individuals and variables for a set of data
____ I can define categorical and quantitative variables
____ I can describe the distribution of a set of data

### While You Read: Key Vocabulary and Concepts

Individuals:

Variable:

Categorical variable:

Quantitative variable:

Distribution:

Inference:

### After You Read: "Where Am I Now?" Check for Understanding

***Concept 1: Individuals and Variables***

Sets of data contain information about individuals. The characteristics of the individuals that are measured are variables. Variables can take on different values for different individuals. It is these values and the variation in them that we will be learning how to study in this course.

***Concept 2: Types of Variables***

Variables can fall into one of two categories: categorical or quantitative. When the characteristic we measure places an individual into one of several groups, we have a categorical variable. When the characteristic we measure results in numerical values for which it makes sense to find an average, we have a quantitative variable. This distinction is important as the methods we use to describe and analyze data will depend upon the type of variable we are studying. One of the first things we will learn is how to display and describe the distribution of a variable.

**Check for Understanding:** ____ *I can identify key characteristics of a set of data.*

Mr. Buckley gathered some information on his class and organized it in a table similar to the one below:

| Student | Gender | ACT Score | Favorite Subject | Grade Point Average |
|---|---|---|---|---|
| James | M | 34 | Statistics | 3.89 |
| Jen | F | 35 | Biology | 3.75 |
| DeAnna | F | 32 | History | 4.00 |
| Jonathan | M | 28 | Literature | 3.00 |
| Doug | M | 33 | Algebra | 2.89 |
| Sharon | F | 30 | Spanish | 3.25 |

1) What individuals does this data set describe?

2) Identify each quantitative and each categorical variable.

3) Describe the distribution of ACT scores.

4) Could we infer from this set of data that students who prefer math and science perform better on the ACT? Explain.

## Section 1.1: Analyzing Categorical Data

### Before You Read: Section Summary

In this section, you will learn how to describe and analyze categorical variables. You will learn how to display data with pie charts and bar graphs as well as how to describe these displays. You will also learn how to analyze the relationship between two categorical variables using marginal and conditional distributions. Finally, you will be introduced to a statistical problem solving strategy called the "4-Step" process.

### "Where Am I Going?"

**Learning Targets:**

_____ I can display categorical data with pie charts or bar graphs
_____ I can distinguish between good and bad graphs
_____ I can construct and interpret two-way tables for categorical variables
_____ I can describe the relationship between two categorical variables using marginal and conditional distributions

### While You Read: Key Vocabulary and Concepts

Frequency table:

Relative frequency table:

Pie chart:

Bar graph:

Two-way table:

Marginal distribution:

Conditional distribution:

Association:

4-Step Process:

**After You Read: "Where Am I Now?"**
**Check for Understanding**

***Concept 1: Displaying Categorical Data***

A frequency table (or a relative frequency table) displays the counts (or percents) of individuals that take on each value of a variable. Tables are sometimes difficult to read and don't always highlight important features of a distribution. Graphical displays of data are much easier to read and often reveal interesting patterns and departures from patterns in the distribution of data. We can use pie charts and bar graphs to display the distribution of categorical variables. When constructing graphical displays, we must be careful not to distort the quantities. Beware of pictographs and watch the scales when displaying or reading graphs!

**Check for Understanding:** ____ *I can display categorical data with pie charts or bar graphs.*

A local business owner was interested in knowing the coffee-shop preferences of her town's residents. According to her survey of 250 residents, 75 preferred "Goodbye Blue Monday," 50 liked "The Ugly Mug," 38 chose "Morning Joe's," 50 said "One Mean Bean," 25 brewed their own coffee, and 12 preferred the national chain.

1) Construct a bar graph to display the data. Describe what you see.

2) Construct a pie chart to display the data. How does this display differ from the bar chart?

***Concept 2: Two-Way Tables***

Bar graphs and pie charts are helpful when analyzing a single categorical variable. However, often we want to explore the relationship between two categorical variables. To do this, we organize our data in a two-way table with a row variable and a column variable. The counts of the individuals in each intersecting category make up the entries in the table.

When exploring a two-way table, you should start by describing each variable separately.

This can be done by describing the marginal distribution of the row or column variable. To explore the relationship between the variables, study the conditional distributions by graphing each distribution of one variable for a fixed value of the other. If there is no association between the two variables, the graphs of the conditional distributions will be similar. If there is an association between the two variables, the graphs of the conditional distributions will be different.

***Strategy: The 4-Step Process***

As we study statistics, we will encounter increasingly complex problems. In this book, we will organize our thinking around four steps:

- **State**: What is the question we are trying to answer?
- **Plan**: What statistical technique does this problem call for?
- **Do**: Make graphs and carry out the necessary calculations.
- **Conclude**: Give a practical conclusion in the context of the situation.

When you encounter a statistical problem, think through these four steps. Always try to state what it is you are trying to answer with the data. Think about what technique would be most appropriate to answer that question. Carry out that technique, showing as much work as possible. Finally, state your conclusion in the context of the problem. Never leave a simple numeric answer!

**Check for Understanding:** ____ *I can describe the relationship between two categorical variables.*

A survey of 1,000 randomly chosen residents of a Minnesota town asked "Where do you prefer to purchase your daily coffee?" The two-way table below shows the responses.

Coffee preference by gender

| **Preference** | **Male** | **Female** | **Total** |
|---|---|---|---|
| National Chain | 95 | 65 | 160 |
| One Mean Bean | 15 | 85 | 100 |
| The Ugly Mug | 145 | 25 | 170 |
| Goodbye Blue Monday | 170 | 90 | 260 |
| Home-brewed | 100 | 160 | 260 |
| Don't drink coffee | 10 | 40 | 50 |
| Total | 535 | 465 | 1000 |

Based on the data, can we conclude that there is an association between gender and coffee preferences? Use appropriate graphical and numerical evidence to support your conclusion. Follow the 4-Step Process.

## Section 1.2: Displaying Quantitative Data with Graphs

**Before You Read: Section Summary**

In this section, you will learn how to display quantitative data. Like categorical data, we are interested in describing the distribution of quantitative data. Dotplots, stemplots, and histograms are helpful tools for revealing the shape, center, and spread of a distribution of quantitative data. These basic plots will come in very handy throughout the course, so be sure to master them in this section!

**"Where Am I Going?"**

**Learning Targets:**

_____ I can construct and interpret a dotplot or stemplot
_____ I can construct and interpret a histogram
_____ I can describe the shape, center, and spread of a distribution
_____ I can identify major departures from the pattern of a distribution (outliers)

**While You Read: Key Vocabulary and Concepts**

Dotplot:

Shape:

Center:

Spread:

Outliers:

Symmetric:

Skewed to the right:

Skewed to the left:

Unimodal:

Bimodal:

Stemplot:

Histogram:

**After You Read: "Where Am I Now?"**
**Check for Understanding**

***Concept 1: Describe the SOCS!***

The reason we construct graphs of quantitative data is so that we can get a better understanding of it. Constructing a plot helps us examine the data and identify its overall pattern. When examining the distribution of a quantitative variable, start by looking for the overall pattern and then describe any major departures from it. Note its shape. Is it symmetric? Is it skewed? How many peaks does it have? What is its center? That is, roughly what value would split the distribution in half? How variable are the data? Are the values bunched up around the center, or are they spread out? Finally, are there any outliers? Do any values fall far away from the rest of the distribution? The answers to each of these questions are very important when describing a set of data. Be sure to address all of them as you explore data sets throughout this course. As you explore data, don't forget your SOCS (Shape, Outliers, Center, Spread)!

***Concept 2: Dotplots and Stemplots***

Dotplots and stemplots are the easiest graphs to construct, especially if you have a small set of data. These plots are helpful for describing distributions because they keep the data intact. That is, you can determine the individual data values directly from the plot. When comparing two sets of data, we can construct back-to-back stemplots or dotplots that share the same scale. Remember, when constructing plots, always label your axes and provide a key!

**Check for Understanding:** ____ *I can construct and interpret dotplots and stemplots.*

A recent study by the Environmental Protection Agency (EPA) measured the gas mileage (miles per gallon) for 30 models of cars. The results are below:

**EPA-Measured MPG for 30 Cars**

| | | | | | | | | | |
|---|---|---|---|---|---|---|---|---|---|
| 36.3 | 32.7 | 40.5 | 36.2 | 38.5 | 36.3 | 41.0 | 37.0 | 37.1 | 39.9 |
| 41.0 | 37.3 | 36.5 | 37.9 | 39.0 | 36.8 | 31.8 | 37.2 | 40.3 | 36.9 |
| 36.7 | 33.6 | 34.2 | 35.1 | 39.7 | 39.3 | 35.8 | 34.5 | 39.5 | 36.9 |

1) Construct a dotplot to display this data and describe the distribution.

2) Construct a stemplot to display this data.

***Concept 2: Histograms***

When dealing with larger sets of data, dotplots and stemplots can be a bit cumbersome and time-consuming to construct. In these cases, it may be easier to construct a histogram. Instead of plotting each value in the data set, a histogram displays the frequency of values that fall within equal-width classes. Be sure not to confuse histograms with bar graphs. Even though they look similar, they describe different types of data!

**Check for Understanding:** ____ *I can construct and interpret a histogram.*

The EPA expanded its study to include a total of 50 car models. The results are below:

**EPA-Measured MPG for 50 Cars**

| | | | | | | | | | |
|---|---|---|---|---|---|---|---|---|---|
| 36.3 | 32.7 | 40.5 | 36.2 | 38.5 | 36.3 | 41.0 | 37.0 | 37.1 | 39.9 |
| 41.0 | 37.3 | 36.5 | 37.9 | 39.0 | 36.8 | 31.8 | 37.2 | 40.3 | 36.9 |
| 36.7 | 33.6 | 34.2 | 35.1 | 39.7 | 39.3 | 35.8 | 34.5 | 39.5 | 36.9 |
| 36.9 | 41.2 | 37.6 | 36.0 | 35.5 | 32.5 | 37.3 | 40.7 | 36.7 | 32.9 |
| 42.1 | 37.5 | 40.0 | 35.6 | 38.8 | 38.4 | 39.0 | 36.7 | 34.8 | 38.1 |

(a) Construct a histogram for these data by hand. Describe the distribution.

(b) Use your calculator to construct a histogram. Do you get the same graph? If not, how can you make your calculator match the histogram you constructed in part (a)?

## Section 1.3: Describing Quantitative Data with Numbers

**Before You Read: Section Summary**

In this section, you will learn how to use numerical summaries to describe the center and spread of a distribution of quantitative data. You will also learn how to identify outliers in a distribution. Being able to accurately describe a distribution and calculate and interpret numerical summaries forms the foundation for our statistical study. By the end of this section, you will want to make sure you are comfortable selecting appropriate numerical summaries, calculating them (by hand or by using technology), and interpreting them for a set of quantitative data.

**"Where Am I Going?"**

**Learning Targets:**

____ I can calculate and interpret measures of center (mean and median)
____ I can calculate and interpret measures of spread (range, IQR, and standard deviation)
____ I can identify outliers using the 1.5 x IQR Rule
____ I can construct and interpret a boxplot
____ I can use appropriate graphs and numerical summaries to compare distributions of quantitative data

**While You Read: Key Vocabulary and Concepts**

Mean:

Median:

Range:

Quartiles:

Interquartile Range (IQR):

Five-Number Summary:

Boxplot:

Standard Deviation:

Variance:

Resistant:

1.5 x IQR Rule:

**After You Read: "Where Am I Now?"**
**Check for Understanding**

***Concept 1: Exploratory Data Analysis***
Statistical tools and ideas can help you examine data in order to describe their main features. This examination is called an ***exploratory data analysis (EDA)***. To organize our exploration, we want to:

- **Examine each variable by itself**...then move on to study relationships among the variables.
- **Always always always always always *plot your data*....**always.
- **Begin with a graph or graphs**...construct and interpret an appropriate graph of the data.
- **Add numeric summaries**...for quantitative data, calculate and interpret appropriate measures of center and measures of spread.
- **Don't forget your SOCS! (S**hape, **O**utliers, **C**enter, **S**pread)...for quantitative data, note these important features!

***Concept 2: Measures of Center***
The mean and median measure center in different ways. While the mean, or average, is the most common measure of center, it is not always the most appropriate. Extreme values can "pull" the mean towards them. The median, or middle value, is resistant to extreme values and is sometimes a better measure of center. In symmetric distributions, the mean and median will be approximately equal. Always consider the shape of the distribution when deciding which measure to use to describe your data!

**Check for Understanding:** ____ *I can calculate and interpret measures of center.*

Consider the following stemplot of the lengths of time (in seconds) it took students to complete a logic puzzle. Use it to answer the following questions. Note: 2|2 = 22 seconds.

| | |
|---|---|
| 1 | 58 |
| 2 | 23 |
| 2 | 677778888999 |
| 3 | 2344 |
| 3 | 68 |
| 4 | 011223 |
| 4 | 6 |
| 5 | 00 |

1) Based only on the plot, how does the mean compare to the median? How do you know?

2) Calculate and interpret the mean. Show your work.

3) Calculate and interpret the median.

4) Which measure of center would be the more appropriate summary of the center of this distribution? Why?

Technology) Use your calculator to find the mean and median for this set of data and verify that you get the same results as (2) and (3).

***Concept 3: Measures of Spread and Boxplots***

Like measures of center, we have several different ways to measure spread. The easiest way to describe the spread of a distribution is to calculate the range (maximum – minimum). However, extreme values can cause this measure of spread to be much greater than the spread of the majority of values. A measure of spread that is resistant to the effect of outliers is the Interquartile range. To find this value, arrange the observations from smallest to largest and determine the median. Then find the median of the lower half of the data. This is the first quartile. The median of the upper half of the data is the third quartile. The distance between the first quartile and third quartile is the Interquartile range (IQR). Not only does the IQR provide a measure of spread, it also provides us with a way to identify outliers. According to the 1.5 x IQR rule, any value that falls more than 1.5 x IQR above the third quartile or below the first quartile is considered an outlier. The minimum, maximum, median, and quartiles make up the "five-number summary." This set of numbers describes the center and spread of a set of quantitative data and leads to a useful display – the boxplot (or box-n-whisker plot).

Another measure of spread that we will use to describe data is the standard deviation. The standard deviation measures roughly the average distance of the observations from their mean. The calculation can be quite time-consuming to do by hand, so we'll rely on technology to provide the standard deviation for us. However, be sure you understand how it is calculated! When describing data numerically, always make sure to note a measure of center and a measure of spread. If you choose the median as your measure of center, it is best to use the IQR to describe the spread. If you choose the mean to describe center, use the standard deviation to measure the spread.

**Check for Understanding:** ____ *I can calculate and interpret measures of spread.*

The length (in pages) of Mrs. Molesky's favorite books are noted below

242 346 314 330 340 322 284 342 368 170 344 318 318 374 332

(a) Use these data to construct a boxplot. Describe the center and spread using the five-number summary.

(b) Calculate and interpret the mean and standard deviation for these data.

## Chapter Summary: Exploring Data

In this chapter, we learned that statistics is the art and science of data. When working with data, it is important to know whether the variables are categorical or quantitative as this will determine the most appropriate displays for the distribution. For categorical data, the display will help us describe the distribution. For quantitative data, the display will help us describe the shape of the distribution and suggest the most appropriate numeric measures of center and spread. Always begin with a graph of the distribution, then move to a numerical description. Which graph, numerical summary, etc. you choose will depend on the context of the situation and the type of data you are dealing with. When exploring quantitative data, we want to be sure to interpret the shape, outliers, center, and spread. Look for an overall pattern to describe your data and note any striking departures from that pattern.

A recurring theme throughout this guide as well as in your textbook is to focus on understanding, not just mechanics. While it may be easy to "plug the data" into your calculator and generate plots and numerical summaries, simply generating graphs and values is not the point of statistics. Rather, you should focus on being able to explain HOW a graph or value is constructed and WHY you would choose a certain display or numeric summary. Get in this habit early...your calculator is a powerful tool, but it can not replace your thinking and communication skills!

### After You Read: "How Can I Close the Gap?"

Complete the vocabulary puzzle, multiple choice questions, and FRAPPY. Check your answers and performance on each of the targets.

| Target | Got It! | Almost There | Needs Some Work |
|---|---|---|---|
| I can identify individuals and variables for a set of data | | | |
| I can define categorical and quantitative variables | | | |
| I can describe the distribution of a set of data | | | |
| I can display categorical data with pie charts or bar graphs | | | |
| I can distinguish between good and bad graphs | | | |
| I can construct and interpret two-way tables for categorical variables | | | |
| I can describe the relationship between two categorical variables using marginal and conditional distributions | | | |
| I can construct and interpret a dotplot or stemplot | | | |
| I can construct and interpret a histogram | | | |
| I can describe the shape, center, and spread of a distribution | | | |
| I can identify major departures from the pattern of a distribution (outliers) | | | |
| I can calculate and interpret measures of center (mean and median) | | | |
| I can calculate and interpret measures of spread (range, IQR, and standard deviation) | | | |
| I can identify outliers using the 1.5 x IQR Rule | | | |
| I can make a boxplot | | | |
| I can use appropriate graphs and numerical summaries to compare distributions of quantitative data | | | |

Did you check “Needs Some Work” for any of the targets? If so, what will you do to address your needs for those targets?

*Learning Plan:*

## Chapter 1 Multiple Choice Practice

**Directions.** *Identify the choice that best completes the statement or answers the question. Check your answers and note your performance when you are finished.*

1. You measure the age (years), weight (pounds), and breed (beagle, golden retriever, pug, or terrier) of 200 dogs. How many variables did you measure?

A. 1
B. 2
C. 3
D. 200
E. 203

2. You open a package of Lucky Charms cereal and count how many there are of each marshmallow shape. The distribution of the variable "marshmallow" is:

A. The shape: star, heart, moon, clover, diamond, horseshoe, balloon.
B. The total number of marshmallows in the package.
C. Seven—the number of different shapes that are in the package.
D. The seven different shapes and how many there are of each.
E. Since "shape" is a categorical variable, it doesn't have a distribution.

3. A review of voter registration records in a small town yielded the following table of the number of males and females registered as Democrat, Republican, or some other affiliation.

| | Male | Female |
|---|---|---|
| Democrat | 300 | 600 |
| Republican | 500 | 300 |
| Other | 200 | 100 |

The proportion of males that are registered as Democrats is:

A. 300
B. 30
C. 0.33
D. 0.30
E. 0.15

4. For a physics course containing 10 students, the maximum point total for the quarter was 200. The point totals for the 10 students are given in the stemplot below.

```
11 | 6 8
12 | 1 4 8
13 | 3 7
14 | 2 6
15 |
16 |
17 | 9
```

Which of the following statements is NOT true?

A. In a symmetric distribution, the mean and the median are equal.
B. About fifty percent of the scores in a distribution are between the first and third quartiles.
C. In a symmetric distribution, the median is halfway between the first and third quartiles.
D. The median is always greater than the mean.
E. The range is the difference between the largest and the smallest observation in the data set.

5. When drawing a histogram it is important to

A. have a separate class interval for each observation to get the most informative plot.
B. make sure the heights of the bars exceed the widths of the class intervals so that the bars are true rectangles.
C. label the vertical axis so the reader can determine the counts or percent in each class interval.
D. leave large gaps between bars. This allows room for comments.
E. scale the vertical axis according to the variable whose distribution you are displaying.

6. A set of data has a mean that is much larger than the median. Which of the following statements is most consistent with this information?

A. The distribution is symmetric.
B. The distribution is skewed left.
C. The distribution is skewed right.
D. The distribution is bimodal.
E. The data set probably has a few low outliers.

7. The following is a boxplot of the birth weights (in ounces) of a sample of 160 infants born in a local hospital.

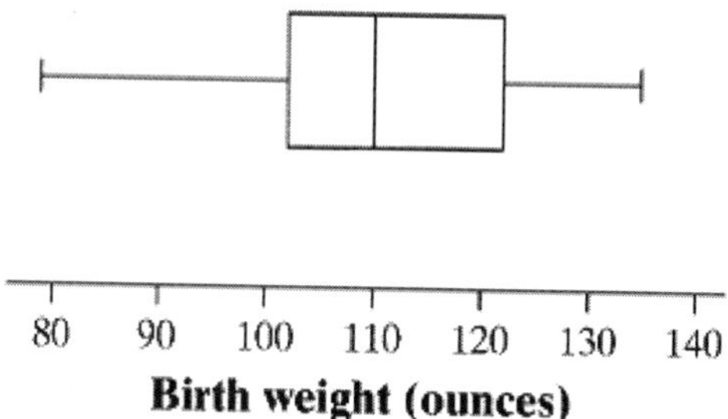


About 40 of the birthweights were below

A. 92 ounces.
B. 102 ounces.
C. 112 ounces.
D. 122 ounces.
E. 132 ounces.

8. A sample of production records for an automobile manufacturer shows the following figures for production per shift:

705 700 690 705

The variance of the sample is approximately

A. 8.66.
B. 7.07.
C. 75.00.
D. 50.00.
E. 20.00.

9. You catch 10 cockroaches in your bedroom and measure their lengths in centimeters. Which of these sets of numerical descriptions are *all* measured in centimeters?

A. median length, variance of lengths, largest length
B. median length, first and third quartiles of lengths
C. mean length, standard deviation of lengths, median length
D. mean length, median length, variance of lengths.
E. both (B) and (C)

10. A policeman records the speeds of cars on a certain section of roadway with a radar gun. The histogram below shows the distribution of speeds for 251 cars.

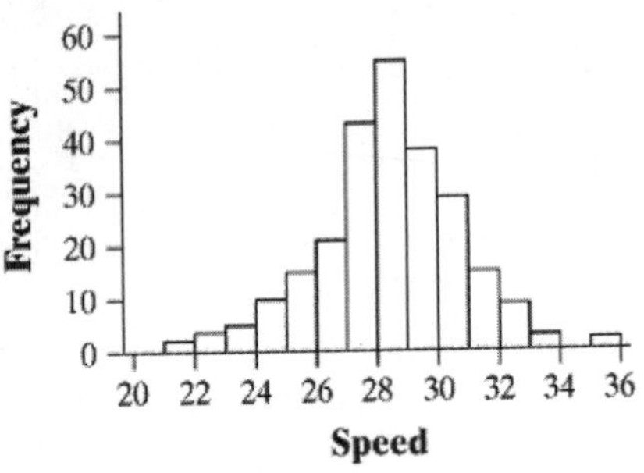


Which of the following measures of center and spread would be the best ones to use when summarizing these data?

A. Mean and interquartile range
B. Mean and standard deviation
C. Median and range
D. Median and standard deviation
E. Median and interquartile range

| Problem | Answer | Concept | Right | Wrong | Simple Mistake? | Need to Study More |
|---|---|---|---|---|---|---|
| 1 | C | Variables | | | | |
| 2 | D | Categorical Variables | | | | |
| 3 | D | Two-way table | | | | |
| 4 | D | Distribution basics | | | | |
| 5 | C | Constructing histograms | | | | |
| 6 | C | Skewed distributions | | | | |
| 7 | B | Interpreting boxplots | | | | |
| 8 | D | Variance | | | | |
| 9 | E | Summary statistics units | | | | |
| 10 | B | Choosing statistics | | | | |

## FRAPPY! Free Response AP Problem, Yay!

The following problem is modeled after actual Advanced Placement Statistics free response questions. Your task is to generate a complete, concise response in 15 minutes. After you generate your response, view two example solutions and determine whether you feel they are "complete," "substantial," "developing" or "minimal." If they are not "complete," what would you suggest to the student who wrote them to increase their score? Finally, you will be provided with a rubric. Score your response and note what, if anything, you would do differently to increase your own score.

SugarBitz candies are packaged in 15 oz. snack-size bags. The back-to-back plot below displays the weights (in ounces) of two samples of SugarBitz bags filled by different filling machines. The weights ranged from 14.1 oz. to 15.9 oz.

| Machine A | Machine | B |
|---:|:---:|:---|
| * | 14 | * |
| | 14 | * * * |
| * * | 14 | * * * * * * * * * * * |
| * * * * | 14 | * * * * * * * |
| * * * * * * * * * * | 14 | * * * * * |
| * * * * * * * * * * * * | 15 | * * * * |
| * * * * * | 15 | * * * |
| * * | 15 | * * |
| | 15 | * |
| | 15 | * |

(a) Compare the distributions of weights of bags packaged by the two machines.

(b) The company wishes to be as consistent as possible when packing its snack bags. Which machine would be *least* likely to produce snack bags of SugarBitz that have a consistent weight? Explain.

(c) Suppose the company filled its bags using the machine you chose in part (b). Which measure of center, mean or median, would be closer to the advertised 15oz.? Explain why you chose this measure.

## How would you score these?

### Student Response 1:

a) Machine A has a slightly higher center than Machine B. Machine B has a much larger range. Machine A is approximately symmetric and Machine B is slightly skewed right. Neither machine has any extreme values.

b) Machine B would be least likely to produce bags containing 15 oz of SugarBitz because it has a much wider range than Machine A.

c) The company should report the mean weight of Machine B. Since the distribution is skewed to the right, the mean will be pulled higher towards the tail. Therefore, the mean will be higher than the median and will be closer to 15.

How would you score this response? Is it substantial? Complete? Developing? Minimal? Is there anything this student could do to earn a better score?

### Student Response 2:

a) Machine A is normal and Machine B is skewed. Both have a single peak and wide ranges.

b) Machine B usually fills bags with about 14.6 oz of candy. Machine A usually fills bags with 15 oz of candy. Machine B is least likely to fill the bags with 15 oz. of candy.

c) The mean because it is about 15.

How would you score this response? Is it substantial? Complete? Developing? Minimal? Is there anything this student could do to earn a better score?

## Scoring Rubric

Use the following rubric to score your response. Each part receives a score of "Essentially Correct," "Partially Correct," or "Incorrect." When you have scored your response, reflect on your understanding of the concepts addressed in this problem. If necessary, note what you would do differently on future questions like this to increase your score.

### Intent of the Question

The goals of this question are (1) to determine your ability to use graphical displays to compare and contrast two distributions and (2) to evaluate your ability to use statistical information to make a decision.

### Solution

**(a)** Both distributions are unimodal (single-peaked). However, Machine A's distribution is roughly symmetric while Machine B's is skewed to the right. The center of the weights for Machine A (median A = about 15) is slightly higher than that of Machine B (median B = about 14.5). There is more variability in the weights produced by Machine B. Machine A has one low value (14.1) that does not fall with the majority of weights. However, it does not appear to be extreme enough to be considered an outlier.

**(b)** Both machines produce bags of varying weight. However, Machine B has a higher variability as evidenced by a wider overall range. Machine B would be least likely to produce a consistent weight for the snack bags.

**(c)** The mean would be closer to the advertised 15 oz. weight. This is because in a skewed distribution, the mean is pulled away from the median in the direction of the tail. In Machine B's distribution, the peak is at about 14.5 oz so we would expect the mean to be higher and closer to 15 oz.

**Scoring**:

Parts (a), (b), and (c) are scored as essentially correct (E), partially correct (P), or incorrect (I).

**Part (a)** is essentially correct if you correctly identify similarities and differences in the shape, center, and spread for the two distributions.
Part (a) is partially correct if you correctly identify similarities and differences in two of the three characteristics for the two distributions.
Part (a) is incorrect if you only identify one similarity or difference of the three characteristics for the two distributions.

**Part (b)** is essentially correct if Machine B is chosen using rationale based on its measure of spread of the packaged weights.
Part (b) is partially correct if B is chosen, but the explanation does not refer to the variability in the weights.
Part (c) is incorrect if B is chosen and no explanation is provided OR if A is chosen.

**Part (c)** is essentially correct if the mean is chosen and the explanation addresses the fact that the mean will be greater than the median in a skewed right distribution.
Part (c) is partially correct if the mean is chosen, but the explanation is incomplete or incorrect.
Part (c) is incorrect if the mean is chosen, but no explanation is given OR if the median is chosen.
NOTE: If Machine A was chosen in part (b) and the solution to part (c) indicates either the mean or median would be appropriate due to the fact that they will be approximately equal in a symmetric, mound-shaped distribution, part (c) should be scored as essentially correct.

4 **Complete Response**
All three parts essentially correct

3 **Substantial Response**
Two parts essentially correct and one part partially correct

2 **Developing Response**
Two parts essentially correct and no parts partially correct
One part essentially correct and two parts partially correct
Three parts partially correct

1 **Minimal Response**
One part essentially correct and one part partially correct
One part essentially correct and no parts partially correct
No parts essentially correct and two parts partially correct

# Chapter 1: Exploring Data

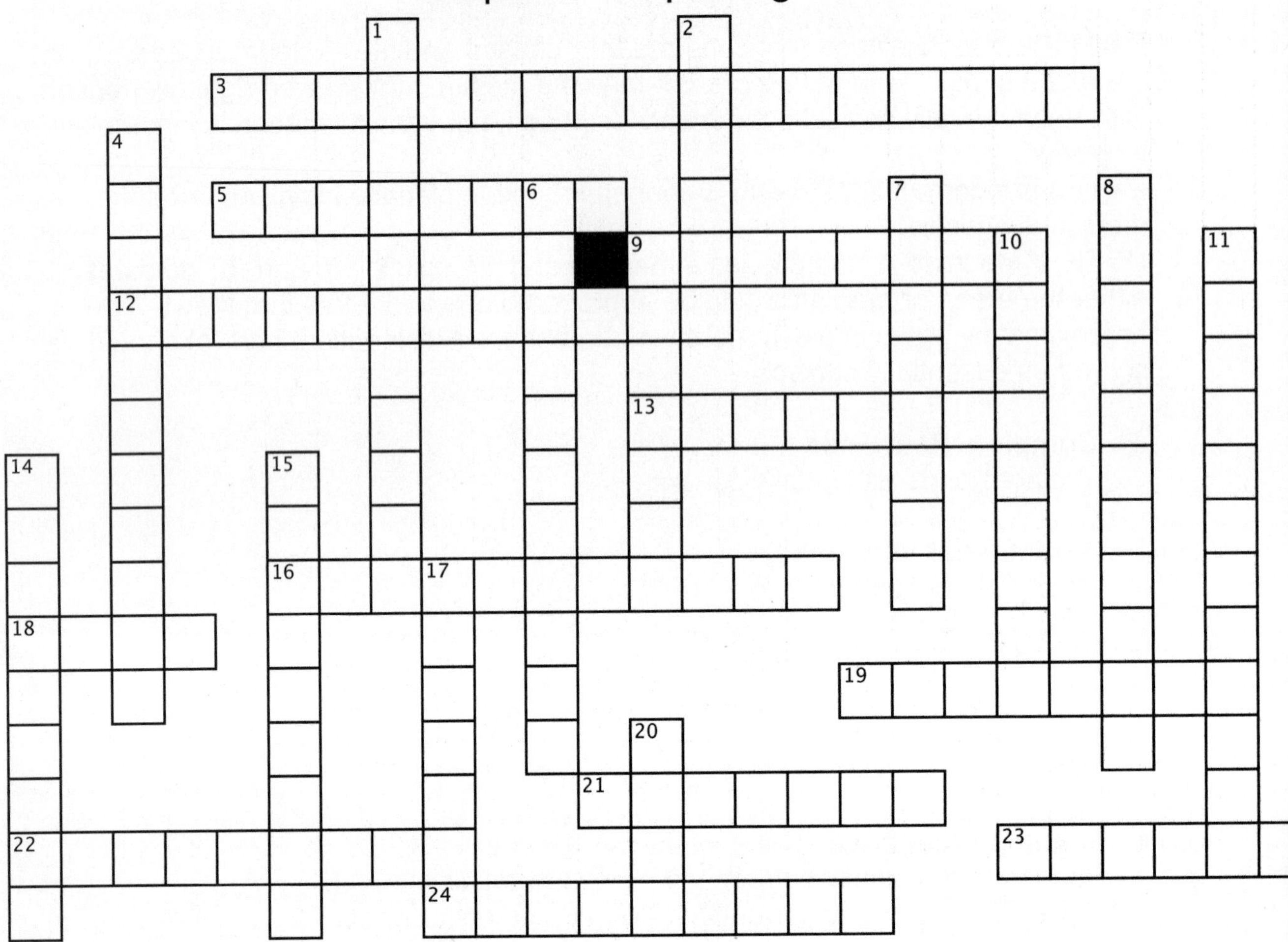

**Across**

3. The average distance of observations from their mean (two words)
5. The average squared distance of the observations from their mean
9. Displays the counts or percents of categories in a categorical variable through differing heights of bars
12. Tells you what values a variable takes and how often it takes these values
13. Displays a categorical variable using slices sized by the counts or percents for the categories
16. When specific values of one variable tend to occur in common with specific values of another
18. A measure of center, also called the average
19. A graphical display of quantitative data that involves splitting the individual values into two components
21. One of the simplest graphs to construct when dealing with a small set of quantitative data
22. Drawing conclusions beyond the data at hand
23. The shape of a distribution if one side of the graph is much longer than the other
24. What we call a measure that is relatively unaffected by extreme observations

**Down**

1. The objects described by a set of data
2. The midpoint of a distribution of quantitative data
4. A _____ distribution describes the distribution of values of a categorical variable among individuals who have a specific v of another variable.
6. A variable that places an individual into one of several group categories
7. A characteristic of an individual that can take different values different individuals
8. When comparing two categorical variables, we can orgainze data in a ___-___ _____.
9. A graphical display of the five-number summary
10. A graphical display of quantitative data that shows the frequ of values in intervals by using bars
11. A variable that takes numerical values for which it makes ser to find an average
14. The shape of a distribution whose right and left sides are approximate mirror images of each other
15. These values lie one-quarter, one-half, and three-quarters of way up the list of quantitative data
17. A value that is at least 1.5 IQRs above the third quartile or be the first quartile
20. When exploring data, don't forget your ____

# Chapter 2: Modeling Distributions of Data

*"It is not knowledge, but the act of learning, not possession but the act of getting there, which grants the greatest enjoyment." Carl Friedrich Gauss*

## Chapter Overview

In Chapter 1, we built a "toolbox" of graphical and numerical tools for describing distributions of data. We now have a clear strategy for exploring data. This chapter introduces you to a key concept in statistics – describing the location of an observation within a distribution. You will learn how to measure position by using percentiles as well as by using a standardized measure based on the mean and standard deviation. • We'll discover that sometimes the overall pattern of a large number of observations is so regular it can be described by a smooth curve. Density curves will be introduced as a model to describe locations without having to rely on every one of the data values. One of the most common density curves, the Normal distribution, will be explored in Section 2.2. You will not only learn the properties of the Normal distributions, but how to perform a number of calculations with them. The concepts introduced in this section will be revisited throughout the course, so be sure to master them in this chapter!

## Sections in this Chapter

**Section 2.1**: Describing Location in a Distribution
**Section 2.2**: Normal Distributions

## Plan Your Learning

Use the following *suggested* guide to help plan your reading and assignments. Note: your teacher may schedule a different pacing. Be sure to follow his or her instructions!

| **Read** | Intro: pp 83-84<br>2.1: pp 85-91 | 2.1: pp 92-104 | 2.2: pp 110-119 | 2.2: pp 119-124 |
|---|---|---|---|---|
| **Do** | 1, 5, 9, 11, 13, 15 | 19, 21,<br>23, 31, 33-38 | 41, 43,<br>45, 47, 49, 51 | 53, 55, 57, 59 |

| **Read** | 2.2: pp 124-130 | Chapter Summary |
|---|---|---|
| **Do** | 63,65,66,68-74 | Multiple Choice<br>FRAPPY! |

## Section 2.1: Describing Location in a Distribution

### Before You Read: Section Summary

In this section, you will learn how to describe the location of an individual in a distribution. You have probably encountered measures of location before through the concept of percentiles. You will learn how to calculate and interpret percentiles as well as how to identify percentiles through a cumulative relative frequency graph. You will also learn a new way to describe location by using the mean and standard deviation. Standardized scores, or *z*-scores, will be introduced as a way to describe location within a distribution and to compare observations from different distributions. Finally, you will be introduced to the concept of a density curve. This smooth curve models the overall shape of a distribution and can be used in place of the actual data to answer questions about the distribution. The concepts you learn in this section will very important throughout the course. Be sure to master the idea of a standardized score!

### "Where Am I Going?"

**Learning Targets:**

_____ I can describe the location of an observation with a percentile

_____ I can interpret a cumulative relative frequency graph

_____ I can describe the location of an observation with a *z*-score and interpret *z*-scores in context

_____ I can describe what happens to the shape, center, and spread of a distribution when data is transformed

_____ I can describe the properties of a density curve

### While You Read: Key Vocabulary and Concepts

percentile:

cumulative relative frequency graph (ogive):

standardized value (*z*-score):

density curve:

effect of adding (or subtracting) a constant:

effect of multiplying (or dividing) by a constant:

**After You Read: "Where Am I Now?"**
**Check for Understanding**

***Concept 1: Measuring Position - Percentiles***

A common way to measure position within a distribution is to tell what percent of observations fall below the value in question. Percentiles are relatively easy to find – especially when our data is presented in order or in a dotplot or stemplot. Cumulative relative frequency graphs, or ogives, provide a graphical tool to find percentiles in a distribution.

**Check for Understanding:** ____ *I can describe the location of an observation in a distribution using percentiles and* ____ *I can interpret cumulative relative frequency graphs.*

The scores on Ms. Chauvet's chapter quiz are displayed in the stemplot and ogive below:

0 | 0 1 2
1 | 2 2 4 8
2 | 1 1 3 4 5 9 9
3 | 0 0 0 3 6
4 | 4 5 7
5 | 0

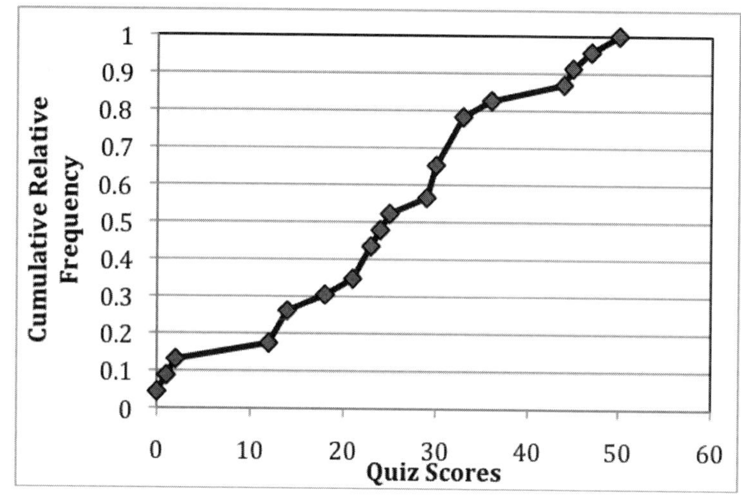


1) James earned a score of 33. Calculate and interpret his percentile.

2) Heather earned a score of 12. Calculate and interpret her percentile.

3) Using the ogive, at what percentile is a score of 38?

4) Using the ogive, estimate and interpret the quartiles of this distribution.

### *Concept 2: Measuring Position – z-Scores*

When describing distributions, we learned that measures of center and spread are both very important characteristics. It follows, then, that measuring the position of an observation in a distribution should take into account both the center and spread of the distribution. After all, saying a particular observation falls 5 points above the average doesn't mean much unless you know how varied the observations are. If the distribution has a small spread, an observation 5 points above average might be an extreme value. However, if the distribution has a large spread, being 5 points above average might not be a big deal. We MUST consider center and spread when describing location. The standardized value, or z-score, of an observation does just that. The z-score tells us how many standard deviations above or below the mean a particular observation falls. This method of describing location not only allows us to describe individuals within a distribution, but also allows us to compare the positions of individuals in different distributions! We will be standardizing values a LOT in this course. Be sure to master the concept now!

**Check for Understanding:** ____ *I can describe the location of an observation in a distribution using z-scores.*

Use the scores from Ms. Chauvet's quiz.

0 | 0 1 2
1 | 2 2 4 8
2 | 1 1 3 4 5 9 9
3 | 0 0 0 3 6
4 | 4 5 7
5 | 0

1) Calculate the mean and standard deviation.

2) Paul earned a 45. Calculate and interpret his z-score.

3) Carl earned a 2. Calculate and interpret his z-score.

4) The scores on Ms. Chauvet's next quiz had a mean of 32 and a standard deviation of 6.5. Paul earned a 47 on this quiz. On which quiz did he perform better relative to the rest of the class? Explain.

***Concept 3: Transforming Data***

When we find *z*-scores, we are actually transforming our data to a standardized scale. That is, we subtract the mean and divide by the standard deviation, converting the observation from its original units to a standardized scale. Sometimes we transform data to switch between measurement units (inches to centimeters, Fahrenheit to Celsius, etc.). When we do this, it is important to know what happens to the center and spread of the transformed distribution. Adding (or subtracting) a constant to each of the observations in a set of data will have an effect on the center of the distribution, but not on the spread or the shape. Multiplying (or dividing) each of the observations by a constant will change the center and the spread of the distribution, but not its shape. Most important, while transforming data in these ways may change the center and spread of a distribution, the locations of individual observations remain unchanged! So, if you had a z-score of 1.5 on a quiz and your teacher decided to double everyone's score and give an additional 5 points, your z-score would STILL be 1.5!

***Concept 4: Density Curves***

You have learned that exploring quantitative data requires making a graph, describing the overall shape, and providing a numerical summary of the center and spread. When we have a large number of observations, we can describe the overall pattern using a smooth curve called a density curve. There are two key features of density curves you need to remember. First, since it is describing the distribution of values it is always on or above the horizontal axis. Second, since it is representing the distribution of all of the values, the area underneath it is exactly 1. Since the area under the entire curve is 1, the area under the curve and between any two values on the horizontal axis is the proportion of observations in the distribution that fall in that interval. One other key point is noted in this section. In a symmetric distribution, the mean and median will be the same. However, in a skewed distribution, the mean will be pulled away from the median in the direction of the tail. Make sure you note that fact; it will come in handy on the AP Exam!

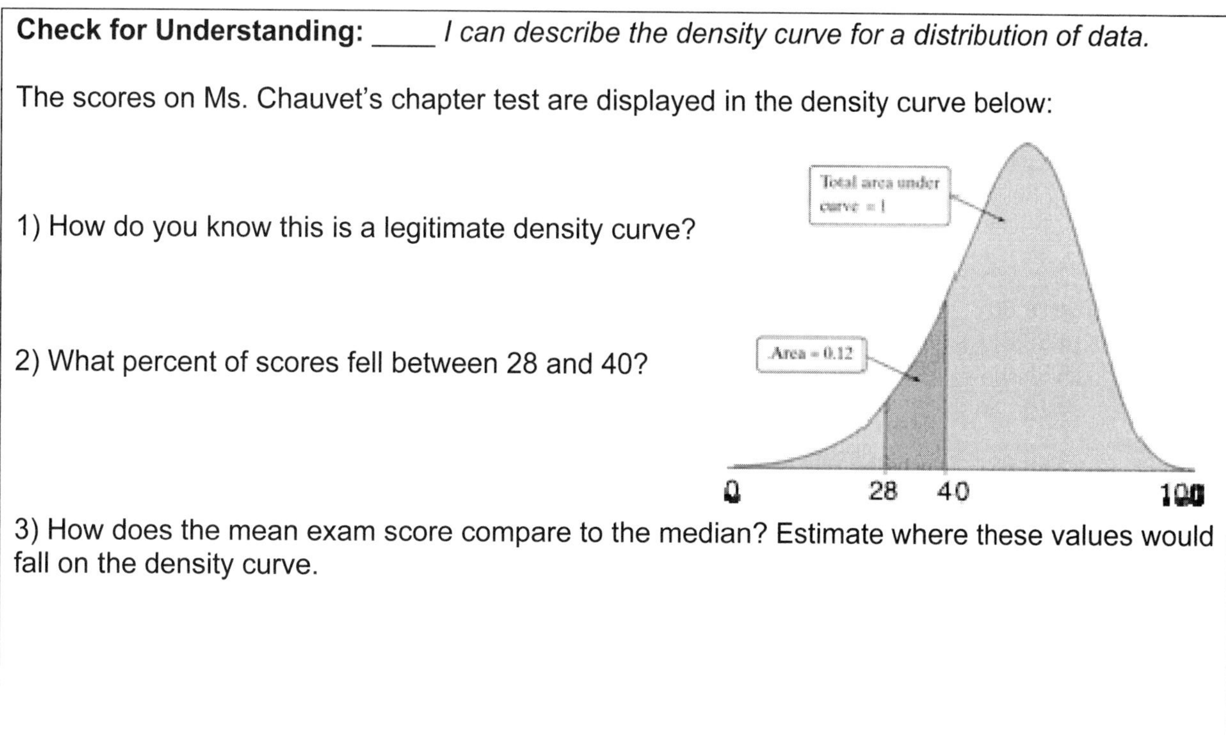

**Check for Understanding:** ____ *I can describe the density curve for a distribution of data.*

The scores on Ms. Chauvet's chapter test are displayed in the density curve below:

1) How do you know this is a legitimate density curve?

2) What percent of scores fell between 28 and 40?



3) How does the mean exam score compare to the median? Estimate where these values would fall on the density curve.

## Section 2.2: Normal Distributions

### Before You Read: Section Summary

As you saw in Section 2.1, density curves are a handy tool for modeling distributions of data. Section 2.2 is devoted entirely to the most common type of density curve – the Normal curve. In this section, you will learn the basic properties of the Normal distributions and how to use them to perform a variety of calculations. You will learn how to determine whether or not a distribution of data can be described as approximately Normal. For distributions that can be described by a Normal curve, you will learn how to determine the proportion of observations that fall into given intervals. You will also learn how to find the value corresponding to a specified percentile in a Normal distribution. Throughout this section, a variety of tools and methods will be presented to help you perform Normal calculations by hand and on your calculator. Remember, whenever you perform a calculation by hand or by calculator, it is important to interpret the results in the context of the situation! Make sure you are comfortable with not only performing the calculations in this section but also in describing what your results mean!

### "Where Am I Going?"

**Learning Targets:**

_____ I can describe and use the 68-95-99.7 Rule
_____ I can find areas and percentiles in the standard Normal distribution
_____ I can perform Normal distribution calculations and interpret their results
_____ I can justify whether or not a distribution of data can be described as Normal

### While You Read: Key Vocabulary and Concepts

Normal distribution:

Normal curve:

68-95-99.7 Rule:

standard Normal distribution:

standard Normal Table:

Normal probability plot:

**After You Read: "Where Am I Now?"**
**Check for Understanding**

***Concept 1: Normal Distributions***

Normal distributions play an important role in statistics. We encounter them a LOT. That's because many distributions of real data and chance outcomes are symmetric, single-peaked, and bell-shaped and can be described by the family of Normal curves. This family of curves shares some important characteristics that help us perform calculations about observations in a distribution of data. First, in distributions that can be described as Normal, the mean is at the center of the Normal curve. Further, almost all of the observations in an approximately Normal distribution will fall within three standard deviations of the mean.

The 68-95-99.7 Rule allows us to be even more specific, noting that in Normal distributions, approximately 68% of the observations will fall within one standard deviation of the mean, approximately 95% of observations will fall within two standard deviations, and approximately 99.7% of observations will fall within three standard deviations. This fact allows us to perform calculations about the approximate proportion of observations in a distribution that fall into certain intervals without even knowing all of the individual observations!

***Concept 2: Normal Distribution Calculations***

As we explore situations in statistics, not all observations of interest will fall one, two, or three standard deviations from the mean. That is, we might be interested in knowing what proportion of observations are at least 1.72 standard deviations above the mean. In cases like this, the 68-95-99.7 rule can help us estimate the proportion, but we'll want to be more exact. Because all Normal distributions share the same properties, the standard Normal table and Normal calculations allow us to perform calculations for *any* observation in an approximately Normal distribution.

To perform a Normal calculation, you should follow the 4-Step Process introduced in Chapter 1. First, express the problem in terms of a variable *x*. Sketch a picture of the distribution and shade the area of interest. Perform calculations by standardizing *x* and then using Table A or your calculator to find the required area under the Normal curve. Finally, be sure to write your conclusion in the context of the problem.

**Check for Understanding:** ____ *I can perform Normal distribution calculations and interpret their results.*

Suppose the scores on all of the quizzes in Ms. Chauvet's class were Normally distributed with mean 83 and standard deviation 5.

1) Sketch the density curve that describes the distribution of scores on the quizzes.

2) What percent of scores are between 78 and 93? Use the 68-95-99.7 rule.

3) A score greater than 90 earns an "A." What percent of quizzes earn an "A"?

4) What percent of scores fall between 71 and 95?

5) What score would place a student in the $20^{th}$ percentile in this class?

***Concept 3: Assessing Normality***

It is important to note that just because a distribution is symmetric, single-peaked, and bell-shaped, it is NOT necessarily Normal. Normal distributions are handy models for some distributions of data. However, before performing Normal calculations, you should be sure the distribution you are exploring really IS Normal! The Normal probability plot of a distribution of data is a helpful tool in assessing Normality of a distribution of data. If the points on a Normal probability plot fall in a relatively straight line, you have evidence that the data are approximately Normal and Normal calculations are justified. If the points taper off or display other systematic departures from a straight line, a Normal model probably isn't appropriate for the distribution. Make sure you are comfortable constructing Normal probability plots on your calculator and interpreting them in the context of the problem!

**Check for Understanding:** ____ *I can justify whether or not a distribution can be considered Normal.*

Use the scores from Ms. Chauvet's first quiz.

0 | 0 1 2
1 | 2 2 4 8
2 | 1 1 3 4 5 9 9
3 | 0 0 0 3 6
4 | 4 5 7
5 | 0

Use graphical and numerical methods to determine whether these data are approximately Normally distributed.

## Chapter Summary: Modeling Distributions of Data

In this chapter, we expanded our toolbox for working with quantitative data. We learned how to describe the location of an individual within a distribution by determining its percentile or by calculating a standardized score ($z$-score) based on the mean and standard deviation of the distribution. We also learned that distributions with a clear overall pattern can be described using a density curve. A common density curve, the Normal curve, is a helpful model for describing many quantitative variables. Knowing how to justify that a distribution is approximately Normally distributed is an important skill. If you can show that a distribution is Normal, you can use the Normal distribution calculations to answer a number of questions about observations within the set of data. The 68-95-99.7 rule and the standard Normal table are both useful tools when performing calculations about observations in Normal distributions.

Like we learned in Chapter 1, you should approach data analysis problems using four steps:

- Graph the data
- Look for an overall pattern and departures from this pattern
- Calculate and interpret numerical summaries
- If the data follow a regular overall pattern, describe the distribution with a smooth curve

The concepts introduced in this chapter will form the basis of much of our study of inference later on in the course. Standardizing data, justifying Normality, and performing Normal calculations are critical skills for statistical inference. Be sure to practice them as you will be using them a LOT!

### After You Read: "How Can I Close the Gap?"

Complete the vocabulary puzzle, multiple choice questions, and FRAPPY. Check your answers and performance on each of the targets.

| Target | Got It! | Almost There | Needs Some Work |
|---|---|---|---|
| I can describe the location of an observation with a percentile | | | |
| I can interpret a cumulative relative frequency graph | | | |
| I can describe the location of an observation with a $z$-score and interpret $z$-scores in context | | | |
| I can describe what happens to the shape, center, and spread of a distribution when data is transformed | | | |
| I can describe and use the 68-95-99.7 Rule | | | |
| I can find areas and percentiles in the standard Normal distribution | | | |
| I can perform Normal distribution calculations and interpret their results | | | |
| I can justify whether or not a distribution of data can be described as Normal | | | |

Did you check "Needs Some Work" for any of the targets? If so, what will you do to address your needs for those targets?

*Learning Plan:*

## Chapter 2 Multiple Choice Practice

**Directions.** *Identify the choice that best completes the statement or answers the question. Check your answers and note your performance when you are finished.*

1. The 16th percentile of a Normally distributed variable has a value of 25 and the 97.5th percentile has a value of 40. Which of the following is the best estimate of the mean and standard deviation of the variable?

| | |
|---|---|
| A. | Mean ≈ 32.5; Standard deviation ≈ 2.5 |
| B. | Mean ≈ 32.5; Standard deviation ≈ 5 |
| C. | Mean ≈ 32.5; Standard deviation ≈ 10 |
| D. | Mean ≈ 30; Standard deviation ≈ 2.5 |
| E. | Mean ≈ 30; Standard deviation ≈ 5 |

2. The proportion of observations from a standard Normal distribution that take values larger than 0.75 is about

| | |
|---|---|
| A. | 0.2266 |
| B. | 0.2500 |
| C. | 0.7704 |
| D. | 0.7764 |
| E. | 0.8023 |

3. The density curve below takes the value 0.5 on the interval $0 < x < 2$ and takes the value 0 everywhere else. What percent of the observations lie between 0.4 and 1.08?

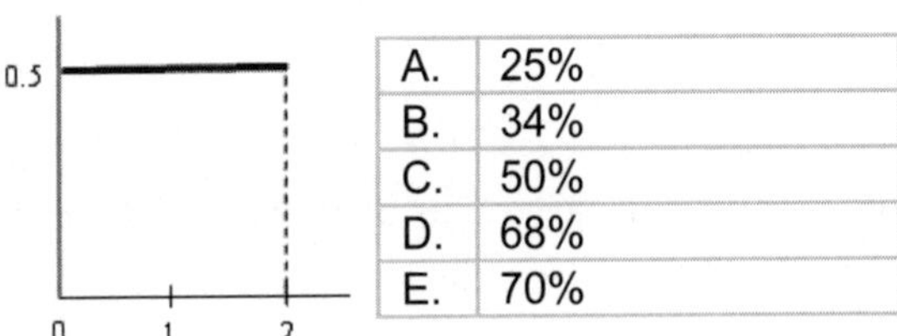


| | |
|---|---|
| A. | 25% |
| B. | 34% |
| C. | 50% |
| D. | 68% |
| E. | 70% |

4. The distribution of the heights of students in a large class is roughly Normal. The average height is 68 inches, and approximately 99.7% of the heights are between 62 and 74 inches. Thus, the variance of the height distribution is approximately equal to

| | |
|---|---|
| A. | 2 |
| B. | 3 |
| C. | 4 |
| D. | 6 |
| E. | 9 |

5. The mean age (at inauguration) of all U.S. Presidents is approximately Normally distributed with a mean of 54.6. Barack Obama was 47 when he was inaugurated, which is the 11th percentile of the distribution. George Washington was 57. What percentile was he in?

| | |
|---|---|
| A. | 6.17 |
| B. | 65.17 |
| C. | 62.92 |
| D. | 34.83 |
| E. | 38.9 |

6. Which of the following statements are false?
I. The standard Normal table can be used with z-scores from any distribution
II. The mean is always equal to the median for any Normal distribution.
III. Every symmetric, bell-shaped distribution is Normal
IV. The area under a Normal curve is always 1, regardless of the mean and standard deviation.

| | |
|---|---|
| A. | I and II |
| B. | I and III |
| C. | II and III |
| D. | III and IV |
| E. | None of the above gives the correct set of true statements. |

7. High school textbooks don't last forever. The lifespan of all high school statistics textbooks is approximately Normally distributed with a mean of 9 years and a standard deviation of 2.5 years. What percentage of the books last more than 10 years?

| | |
|---|---|
| A. | 11.5% |
| B. | 34.5% |
| C. | 65.5% |
| D. | 69% |
| E. | 84.5% |

8. The distribution of the time it takes for different people to solve their Strive for a Five chapter crossword puzzle is strongly skewed to the right, with a mean of 10 minutes and a standard deviation of 2 minutes. The distribution of *z*-scores for those times is

| | |
|---|---|
| A. | Normally distributed, with mean 10 and standard deviation 2. |
| B. | Skewed to the right, with mean 10 and standard deviation 2. |
| C. | Normally distributed, with mean 0 and standard deviation 1. |
| D. | Skewed to the right, with mean 0 and standard deviation 1. |
| E. | Skewed to the right, but the mean and standard deviation cannot be determined without more information. |

9. The cumulative relative frequency graph below shows the distribution of lengths (in centimeters) of fingerlings at a fish hatchery. The third quartile for this distribution is approximately:

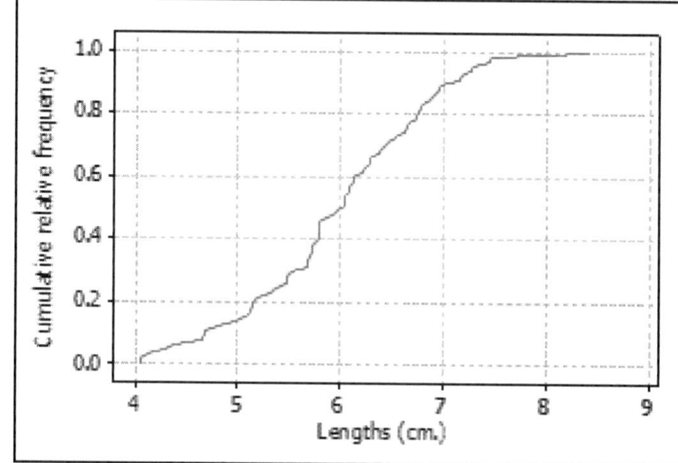


| | |
|---|---|
| A. | 6.7 cm |
| B. | 7 cm |
| C. | 6 cm |
| D. | 5.5 cm |
| E. | 7.5 cm |

10. The plot shown below is a Normal probability plot for a set of test scores. Which statement is true for these data?

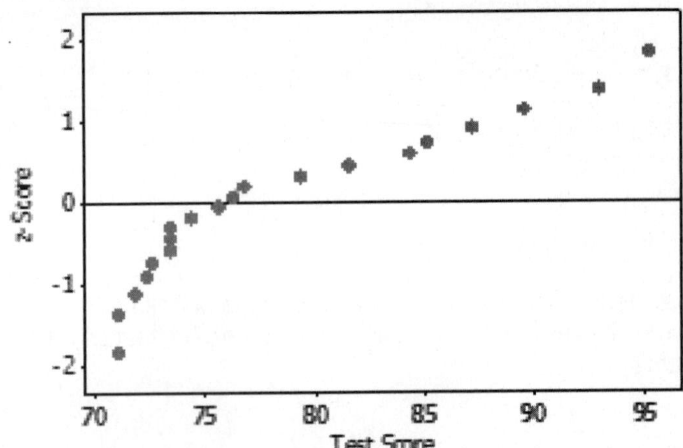


| | |
|---|---|
| A. | The distribution is clearly Normal. |
| B. | The distribution is approximately Normal. |
| C. | The distribution appears to be skewed. |
| D. | The distribution appears to be uniform. |
| E. | There is insufficient information to determine the shape of the distribution. |

| Problem | Answer | Concept | Right | Wrong | Simple Mistake? | Need to Study More |
|---|---|---|---|---|---|---|
| 1 | E | 68-95-99.7 Rule | | | | |
| 2 | A | Standard Normal Table | | | | |
| 3 | B | Area under a Density Curve | | | | |
| 4 | C | 68-95-99.7 Rule and Variance | | | | |
| 5 | B | Standard Normal Calculations | | | | |
| 6 | B | Properties of Normal Distributions | | | | |
| 7 | B | Standard Normal Calculations | | | | |
| 8 | D | Standardized Scores | | | | |
| 9 | A | Cumulative Relative Frequency Graph | | | | |
| 10 | C | Normal Probability Plots | | | | |

## FRAPPY! Free Response AP Problem, Yay!

The following problem is modeled after actual Advanced Placement Statistics free response questions. Your task is to generate a complete, concise response in 15 minutes. After you generate your response, view two example solutions and determine whether you feel they are "complete," "substantial," "developing," or "minimal." If they are not "complete," what would you suggest to the student who wrote them to increase their score? Finally, you will be provided with a rubric. Score your response and note what, if anything, you would do differently to increase your own score.

Final exam grades are determined by the percent correct on the exam. A teacher's records indicate the performance on the exam is Normally distributed with mean 82 and standard deviation 5. The grades on her exam are assigned using the scale below.

| Grade | Percent Correct |
|---|---|
| A | 94 ≤ percent ≤ 100 |
| B | 85 ≤ percent < 94 |
| C | 76 ≤ percent < 85 |
| D | 65 ≤ percent < 76 |
| F | 0 ≤ percent < 65 |

(a) Use a sketch of a Normal distribution to illustrate the proportion of students who would earn a B. Calculate this proportion.

(b) Students who earn a B, C, or D, are considered to "meet standards." Based on this grading scale, what percent of students will receive a score that places them in a category other than "meets standards"?

(c) What grade would the student who scored at the $25^{th}$ percentile earn on this chapter? Justify your answer.

## How would you score these?

### Student Response 1:

a) P(B) = 0.9918 – 0.7257 = 0.2661. 26.61 % of students earn a B.
b) P(does not meet standards) = P(F) = P(z<3.4) = 0.0003
c) z = 0.25 so the score would be 82 + 0.25(5) = 83.25

How would you score this response? Is it substantial? Complete? Developing? Minimal? Is there anything this student could do to earn a better score?

### Student Response 2:

a) P(B) = P( 0.6 ≤ z < 2.4) = 0.9793 – 0.5239 = 0.4554
b) P(A or F) = P(z > 2.4) + P(z < 3.4) = 0.0085
c) A z-score of -0.6745 corresponds to the 25$^{th}$ percentile. So, the score would be 82+ (-0.6475)(5) = 78.8.

How would you score this response? Is it substantial? Complete? Developing? Minimal? Is there anything this student could do to earn a better score?

## Scoring Rubric

Use the following rubric to score your response. Each part receives a score of "Essentially Correct," "Partially Correct," or "Incorrect." When you have scored your response, reflect on your understanding of the concepts addressed in this problem. If necessary, note what you would do differently on future questions like this to increase your score.

### Intent of the Question

The goal of this question is to determine your ability to perform and interpret Normal calculations.

### Solution

**(a)**

85 ≤ percent < 94 )
= P( (85-82)/5 ≤ z < (94-82)/5 )
= P( 0.6 ≤ z < 2.4)
= 0.9918 – 0.7257
= 0.2661

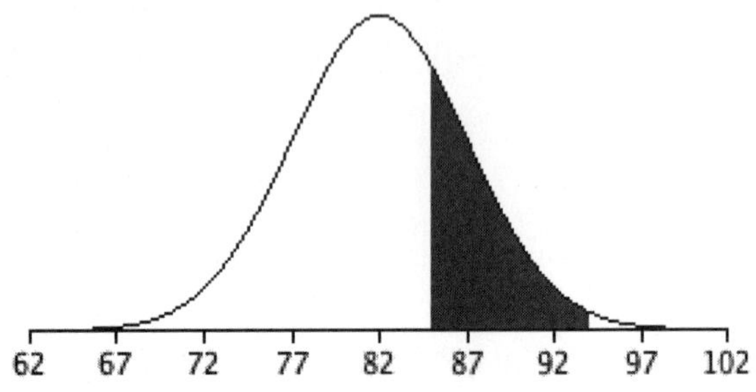


**(b)** P(A or F) = P(A) + P(F)
= P(z ≥ (94-82)/5) + P(z < (65-82)/5)
= P(z ≥ 2.4) + P(z < 3.4)
= 0.0082 + 0.0003
= 0.0085

**(c)** A z-score of -0.6745 corresponds to the $25^{th}$ percentile.
x = mean + z(std dev)
x = 82+ (-0.6475)(5)
x = 78.8

### Scoring:

Parts (a), (b), and (c) are scored as essentially correct (E), partially correct (P), or incorrect (I).

**Part (a)** is essentially correct if the response (1) recognizes the need to look at grades of A and F and (2) correctly computes the tail probabilities and adds them together.

Part (a) is partially correct if the response
Considers only an A or an F and calculates the corresponding tail area correctly
Recognizes the need to look at A and F but only calculates one of the tail areas correctly
Approximates the probabilities using the Empirical rule
Computes the proportion that will "meet standards"
States the correct answer 0.0085 without supporting work

**Part (b)** is essentially correct if (1) the appropriate probability is illustrated using a labeled Normal curve and (2) the proportion is correctly computed.
Part (b) is partially correct if only one of the above elements is correct.

**Part (c)** is essentially correct if the student recognizes the situation as binomial and identifies p from part (b) and shows work to calculate the correct probability.
Part (c) is partially correct if the student recognizes the situation as binomial and identifies p, but does not compute the correct probability OR gives the correct probability but does not show work.

NOTE: If the student makes an error in part (b) and correctly uses that probability in part (c) to compute a reasonable probability, part (c) is essentially correct.

**4 Complete Response**
All three parts essentially correct

**3 Substantial Response**
Two parts essentially correct and one part partially correct

**2 Developing Response**
Two parts essentially correct and no parts partially correct
One part essentially correct and two parts partially correct
Three parts partially correct

**1 Minimal Response**
One part essentially correct and one part partially correct
One part essentially correct and no parts partially correct
No parts essentially correct and two parts partially correct

# Chapter 2: Modeling Distributions of Data

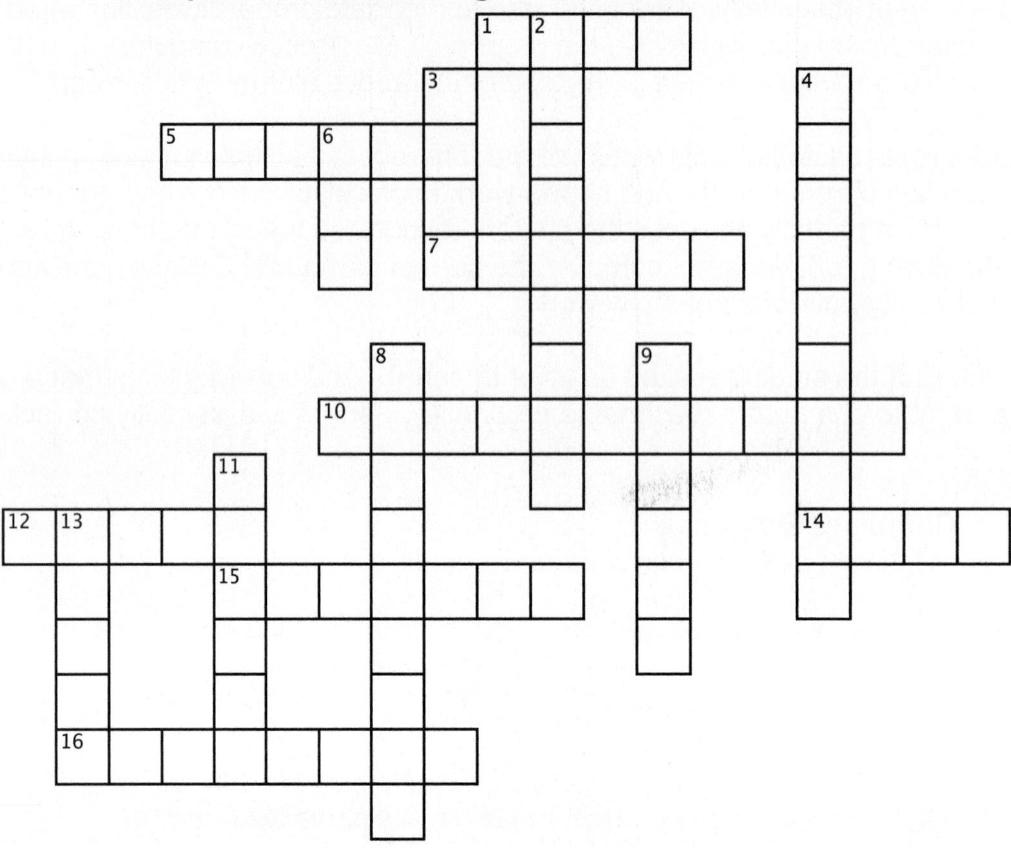

**Across**

1. The balance point of a density curve, if it were made of solid material
5. The standardized value of an observation
7. These common density curves are symmetric and bell-shaped
10. A Normal _____ plot provides a good assessment of whether a data set is approximately Normally distributed
12. Another name for a cumulative relative frequency graph
14. The standard Normal table tells us the area under the standard Normal curve to the ___ of z
15. A ___ curve is a smooth curve that can be used to model a distribution
16. This Normal distribution has mean 0 and standard deviation 1

**Down**

2. The ____ rule is also known as the 68-95-99.7 rule for Normal distributions
3. To standardize a value, subtract the ___ and divide by the standard deviation
4. The value with p percent of the observations less than it
6. The area under any density curve is always equal to
8. We ___ data when we change each value by adding a consta[nt] and/or multiplying by a constant.
9. If a Normal probability plot shows a _____ pattern, the data a[re] approximately Normal
11. The point that divides the area under a density curve in half
13. This mathematician first applied Normal curves to data to err[ors] made by astronomers and surveyors

# Chapter 3: Describing Relationships

*"You can only predict things after they've happened."* Eugene Ionesco

## Chapter Overview

Our statistics toolbox now contains a variety of ways to explore a single quantitative variable. Further, we have learned ways to explore one or more categorical variables. Often in our studies, though, we will need to explore and describe the relationship between two quantitative variables. In this chapter, we will learn how to analyze patterns in "bivariate" relationships by plotting them and calculating summary statistics about them. Further, we will learn how to describe them using mathematical models that can be used to make predictions based on the relationship between the variables. Investigating the relationship between two variables is a key component of statistical study and is the final skill necessary for our data exploration toolbox. Be sure to master the concepts and methods in this chapter!

## Sections in this Chapter



## Plan Your Learning

Use the following *suggested* guide to help plan your reading and assignments. Note: your teacher may schedule a different pacing. Be sure to follow his or her instructions!

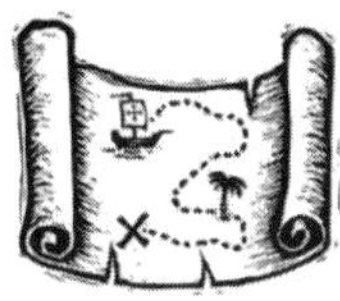

| **Read** | Intro: pp 142-143<br>3.1: pp 143-149 | 3.1: pp 150-157 | 3.2:<br>pp 164-167 | 3.2: pp 168-174 |
|---|---|---|---|---|
| **Do** | 1, 5, 7, 11, 13, | 14-18, 21, 26 | 27-32,<br>35, 37, 39, 41 | 43, 45, 47, 53 |

| **Read** | 3.2: pp 174-181 | 3.2: pp 181-190 | Chapter Summary |
|---|---|---|---|
| **Do** | 49, 54, 56, 58-61 | 63, 65, 68, 69,<br>71-78 | Multiple Choice<br>FRAPPY! |

## Section 3.1: Scatterplots and Correlation

### Before You Read: Section Summary

Many statistical studies examine more than one variable. So far, we have learned methods to graph and describe relationships between categorical variables. In this chapter, we'll learn that the approach to data analysis that we learned for a single quantitative variable can also be applied to explore the relationship between two quantitative variables. That is, we'll learn how to plot our data and add numerical summaries. We'll then learn how to describe the overall patterns and departures from patterns that we see. Finally, we'll learn how to create a mathematical model to describe the overall pattern. This section will focus primarily on displaying the relationship between two quantitative variables and describing its form, direction, and strength. Like the previous chapters, you will find that technology can be used to do most of the difficult calculations. However, be sure you understand *how* the calculator is determining its results and *what* those results mean!

### "Where Am I Going?"

**Learning Targets:**

- _____ I can identify explanatory and response variables in bivariate situations
- _____ I can construct and interpret a scatterplot to display a bivariate relationship
- _____ I can describe the direction, form, and strength of the pattern in a scatterplot
- _____ I can calculate and interpret correlation
- _____ I can identify outliers in a scatterplot and explain their effects on correlation

### While You Read: Key Vocabulary and Concepts

response variable:

explanatory variable:

scatterplot:

direction, form, strength:

outlier:

positive association:

negative association:

correlation *r:*

**After You Read: "Where Am I Now?"**
**Check for Understanding**

***Concept 1: Explanatory vs. Response Variables***

The purpose of many studies of bivariate relationships is to develop a model so that we can use one variable to make a prediction for the other. Because of that, it is important to clearly identify which variable in a situation is explanatory and which is the response. The explanatory variable is the one we think explains the relationship or "predicts" changes in the response variable. It is important to know the difference as identifying the explanatory and response variable will determine how we display the data and how we calculate a summary model. Note: you may have learned about independent and dependent variables in an earlier math or science class. Those are just different names for explanatory and response variables. We'll avoid using independent and dependent, though, because those terms have a different meaning later in the course.

**Check for Understanding:** ____ *I can identify explanatory and response variables.*

Identify the explanatory and response variable in the following situations:

How does stress affect your test performance? In a recent study, researchers studied students' test anxiety and subsequent performance on a standardized test.

Is brain size related to memory? A 1995 study measured the volume of each subject's hippocampus and then administered a short verbal retention assessment.

***Concept 2: Scatterplots***

Like we learned in chapters 1 and 2, plotting data should always be our first step in a data exploration. The most useful graph for exploring bivariate relationships is the scatterplot. Making a scatterplot is pretty easy. 1) Determine which variable goes on which axis. (Hint: eXplanatory goes on the x-axis!) 2) Label and scale the axes. 3) Plot individual data values. Once you have the scatterplot constructed, take some time to describe what you see. What is the overall form of the relationship? Is it linear? Nonlinear? What direction does the relationship take? Is it positive or negative? How strong is the relationship? Do the points follow the pattern closely, or are they widely scattered? Finally, are there any outliers?

**Check for Understanding:** ____ *I can construct and interpret a scatterplot.* ____ *I can describe the form, direction, and strength of a relationship displayed in a scatterplot.*

Use the following data to construct a scatterplot and describe the form, direction, and strength of the relationship between anxiety and exam performance. Note: Higher anxiety scores indicate higher levels of test anxiety.

| Anxiety | 23 | 14 | 14 | 0 | 7 | 20 | 20 | 15 | 21 | 4 |
|---|---|---|---|---|---|---|---|---|---|---|
| Exam Score | 43 | 59 | 48 | 77 | 50 | 52 | 46 | 60 | 51 | 70 |

Scatterplot:

***Concept 3: Measuring Linear Relationships - Correlation***

Scatterplots are great tools for displaying the direction, form, and strength of the relationship between two quantitative variables. Often, we will want to know whether or not the relationship is linear and, if so, how strong the linear relationship is. However, our eyes aren't the most accurate judges of the strength of linear relationships. Correlation $r$ provides a numerical summary of the strength of the linear relationship that can be easily interpreted. Some key points to remember about $r$ include the fact that it is always a number between -1 and 1. Perfect linear relationships are defined when $r = 1$ or $r = -1$. Positive relationships have a positive correlation and vice versa. Finally, the closer $|r|$ is to 1, the stronger the linear relationship between the quantitative variables. Note, however, just because the linear relationship is strong, it is possible that curvature still exists! Scatterplots that display little to no pattern will have a correlation close to 0.

**Check for Understanding:** ____ *I can calculate and interpret correlation.*

Use the data from the anxiety vs. exam score example to calculate and interpret the correlation coefficient. What does this value tell you about the relationship?

## Section 3.2: Least-Squares Regression

### Before You Read: Section Summary

In the last section, we learned that we can display the relationship between two quantitative variables using a scatterplot. Further, we can use a scatterplot to describe the direction, form, and strength of the relationship. The correlation coefficient $r$ allows us to further describe the situation by telling us how strong the *linear* relationship between the variables is. In this section, we'll learn how to summarize the overall pattern of a linear relationship by finding the equation of the least squares regression line. This line can be used to not only model the linear relationship, but also to make predictions based on the overall pattern. Like correlation, our calculator will do most of the work for us. Your job is to be able to interpret and apply the results!

### "Where Am I Going?"

**Learning Targets:**

- ____ I can construct or identify the equation a least-squares regression line
- ____ I can interpret the slope and *y*-intercept of a least-squares regression line
- ____ I can calculate and interpret residuals
- ____ I can construct and interpret residual plots
- ____ I can explain the dangers of extrapolation
- ____ I can use the least-squares regression line to predict values of the response variable
- ____ I can use the standard deviation of the residuals to assess how well the line fits the data
- ____ I can use $r^2$ to assess how well the line fits the data

### While You Read: Key Vocabulary and Concepts

regression line:

predicted value:

slope:

y-intercept:

extrapolation:

least-squares regression line:

residual:

residual plot:

standard deviation of the residuals, *s:*

coefficient of determination $r^2$:

outliers and influential points:

correlation vs. causation:

**After You Read: "Where Am I Now?"**
**Check for Understanding**

***Concept 1: Least-Squares Regression Line***
The main concept in this section is that of the least-squares regression line. When a scatterplot suggests a linear relationship between quantitative explanatory and response variables, we can summarize the pattern by "fitting" a line to the points. This line can then be used to predict values of the response variable for given values of the explanatory variable. Note, however, we should use caution not to make predictions too far outside of our observed x-values. Extrapolation can be dangerous, as we don't know whether or not the pattern continues outside our observations!

The equation of the least-squares regression line can be calculated by hand, if you know the mean and standard deviation of the variables and the correlation *r*. However, you might want to rely on technology to provide the equation for you. Focus your energy on interpreting the slope and *y*-intercept in context and on using the model to make predictions. The y-intercept is often meaningless in the context of our situations. It tells us what response value we'd predict to see if our explanatory value was zero. The slope is the key value of interest in describing the relationship between two quantitative variables. It tells us how much of an increase (or decrease) we expect to see, on average, in our predicted y-values for each one-unit increase in our x-values. Get familiar with that concept as we will see it again in future chapters!

By the end of this section, you should be able to construct and interpret a least-squares regression model, justify its use, and use it to make predictions. Be sure to focus on interpreting the different components of the model!

***Concept 1: Regression Lines-Prediction and Extrapolation***
A regression line is a line the models the data. That is, it summarizes the overall pattern and provides an equation that represents the relationship between our explanatory and response variable. This equation can be used to predict the response for a given value of the explanatory

variable. Use caution not to extrapolate when making predictions, though, as we do not know if the relationship between the variables extends far beyond the observed values of *x*!

***Concept 2: Least-Squares Regression Line***

Chances are any scatterplot you construct or encounter will not display a perfectly straight line. In most cases, the observed points will be, well, scattered. Since most of our observed relationships are not perfectly linear, predictions of *y* made from our regression line will often be different than observed *y* values, resulting in a prediction error. That is, there will be some amount of vertical distance between the regression line and the observed value. This vertical difference (observed *y* – predicted *y*) is called a residual. The regression line that "best fits" our observed data is the one that minimizes the squared residuals. This "line of best fit" that minimizes that prediction error is called the least-squares regression line.

Familiarize yourself with the formulas that can be used to determine the slope and intercept of the least-squares regression line. We will rely on technology to generate this equation, but you should recognize that we can construct the equation by hand given the mean and standard deviation of *x* and *y* as well as the correlation *r* between them.

Once you have the equation of the least-squares regression line, you should be able to interpret it and use it. The most important feature to note when interpreting is the slope. You should be able to explain what the slope means in the context of the variables you are analyzing. That is, the slope represents the expected change in the predicted *y* value for each one-unit increase of the *x* value. Be sure to get familiar with this interpretation as you may be asked to provide it on the AP Exam!

**Check for Understanding:** ____ *I can construct, interpret, and apply the least-squares regression line.*

Using the following data, determine the least-squares regression line to predict exam scores from anxiety scores. Note: Higher anxiety scores indicate higher levels of test anxiety.

| Anxiety | 23 | 14 | 14 | 0 | 7 | 20 | 20 | 15 | 21 | 4 |
|---|---|---|---|---|---|---|---|---|---|---|
| Exam Score | 43 | 59 | 48 | 77 | 50 | 52 | 46 | 60 | 51 | 70 |

a) What is the equation of the least-squares regression line?

b) Interpret the slope of the least-squares regression line in the context of the situation.

c) What exam score can we predict for an anxiety score of 15?

d) What is the residual for an anxiety score of 15?

e) Would you use your least-squares regression line to predict an exam score for a person who had an anxiety score of 35? Why or why not?

***Concept 3: Assessing How Well the Least-Squares Regression Line Fits the Data***

In Section 3.1, we learned that our eyes aren't always the best judge of linear relationships. While correlation *r* gives us a better understanding of the strength of the linear relationship, we still need to assess how well the least-squares regression line fits the observed data. If it fits well, it may be a useful prediction tool. If it doesn't fit well, we may want to search for a model that fits it better.

One way to assess how well the least-squares regression line fits our data is to make a residual plot. Plotting the residuals gives us more information about the relationship between quantitative variables and helps us assess how well a linear model fits the data. If the residual plot displays a pattern, a better (perhaps nonlinear) model might exist!

We can also assess the fit of the least-squares regression line by interpreting the coefficient of determination $r^2$. $r^2$ is a measure of how well the regression model explains the response. Specifically, it is interpreted as the fraction of variation in the values of *y* that is explained by the least-squares regression line of *y* on *x*. For example, if $r^2 = 0.82$, we can say that 82% of the variation in y is due to the linear relationship between *y* and *x*. 18% is due to factors other than *x*.

**Check for Understanding:** ____ *I can assess how well the least-squares regression line fits the data.*

Consider the equation of the least squares regression line of exam score on anxiety.

1) Construct and interpret the residual plot for the least-squares regression line.

2) What is the value of *r*? What is the value of $r^2$? Interpret each of these in the context of the problem.

***Concept 4: Interpreting Computer Regression Output***

As noted already, we will often rely on technology to generate the equation of the least-squares regression line. You are probably familiar with using your calculator to produce the equation. Make sure you can also interpret computer output to identify the slope and intercept of the regression line as well as other important values such as correlation and the coefficient of determination. There is a strong possibility you will need to read computer output on the AP Exam!

**Check for Understanding:** ____ *I can construct or identify the equation of a least squares regression line.*

A study was performed to determine the effect of temperature on a pond's algae level. Temperature was measured in degrees F, and algae level was measured in parts per million. Consider the computer output below.

```
Predictor  Coef      Stdev    t-ratio   p
Constant  42.8477   5.750     77.40     0.000
Temp       0.47620  0.5911    13.70     0.000
s = 0.4224          R-sq= 91.7%   R-sq(adj)=91.2%
```

1) Write the equation of the least squares regression line. Identify any variables used.

2) Interpret the slope of the least-squares regression line.

3) Identify and interpret the correlation coefficient.

4) Identify and interpret the standard deviation of the residuals.

## Chapter Summary: Modeling Distributions of Data

In this chapter, we expanded our toolbox for working with quantitative data. We learned how to analyze and describe the relationship between two quantitative variables. Using scatterplots, we can display the relationship and describe the direction, strength, and form of the overall pattern. Correlation provides a numerical summary of the strength of the linear relationship between the variables and the equation of the least-squares regression line provides a model that can be used to make predictions. Residual plots, the standard deviation of the residuals, and the coefficient of determination help us assess the fit of the least-squares regression line and may suggest whether or not a linear model is appropriate. Finally, we learned that outliers and influential points can affect our interpretations and regression results. Just like we did with a single quantitative variable, we should be able to identify departures from the overall pattern and explain their influence on our analysis.

Perhaps the most important note for this chapter, though, is that while we now have some tools to help us describe the relationship between two quantitative variables, correlation does not always imply causation!

### After You Read: "How Can I Close the Gap?"

Complete the vocabulary puzzle, multiple choice questions, and FRAPPY. Check your answers and your performance on each of the targets.

| Target | Got It! | Almost There | Needs Some Work |
|---|---|---|---|
| I can identify explanatory and response variables in bivariate situations | | | |
| I can construct and interpret a scatterplot to display a bivariate relationship | | | |
| I can describe the direction, form, and strength of the pattern in a scatterplot | | | |
| I can calculate and interpret correlation | | | |
| I can identify outliers in a scatterplot and explain their effects on correlation | | | |
| I can construct or identify the equation of a least-squares regression line | | | |
| I can interpret the slope and *y*-intercept of a least-squares regression line | | | |
| I can calculate and interpret residuals | | | |
| I can construct and interpret residual plots | | | |
| I can explain the dangers of extrapolation | | | |
| I can use the least-squares regression line to predict values of the response variable | | | |
| I can use the standard deviation of the residuals to assess how well the line fits the data | | | |
| I can use $r^2$ to assess how well the line fits the data | | | |

Did you check "Needs Some Work" for any of the targets? If so, what will you do to address your needs for those targets?

*Learning Plan:*

## Chapter 3 Multiple Choice Practice

**Directions.** *Identify the choice that best completes the statement or answers the question. Check your answers and note your performance when you are finished.*

1. A study is conducted to determine if one can predict the academic performance of a first year college student based on their high school grade point average. The explanatory variable in this study is

| | |
|---|---|
| A. | academic performance of the first year student. |
| B. | grade point average. |
| C. | the experimenter. |
| D. | number of credits the student is taking. |
| E. | the college. |

2. If two variables are positively associated, then

| | |
|---|---|
| A. | larger values of one variable are associated with larger values of the other. |
| B. | larger values of one variable are associated with smaller values of the other. |
| C. | smaller values of one variable are associated with larger values of the other. |
| D. | smaller values of one variable are associated with both larger or smaller values of the other. |
| E. | there is no pattern in the relationship between the two variables. |

3. The correlation coefficient measures

| | |
|---|---|
| A. | whether there is a relationship between two variables. |
| B. | the strength of the relationship between two quantitative variables. |
| C. | whether or not a scatterplot shows an interesting pattern. |
| D. | whether a cause and effect relation exists between two variables. |
| E. | the strength of the linear relationship between two quantitative variables. |

4. Consider the following scatterplot, which describes the relationship between stopping distance (in feet) and air temperature (in degrees Centigrade) for a certain 2,000-pound car travelling 40 mph.

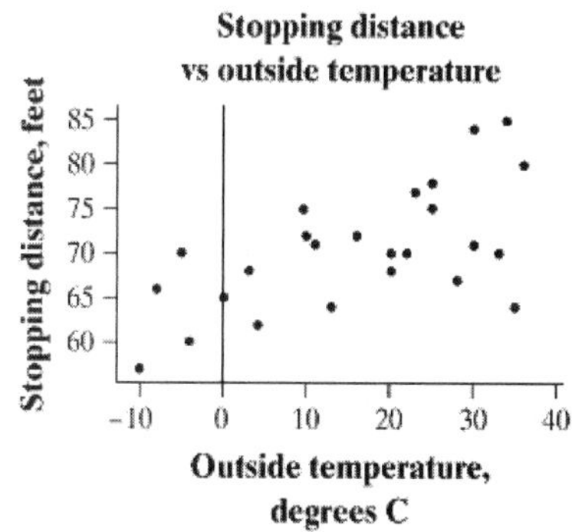


Do these data provide strong evidence that warmer temperatures actually *cause* a greater stopping distance?

| | |
|---|---|
| A. | Yes. The strong straight-line association in the plot shows that temperature has a strong effect on stopping distance. |
| B. | No. $r \neq +1$ |
| C. | No. We can't be sure the temperature is responsible for the difference in stopping distances. |
| D. | No. The plot shows that differences among stopping distances are not large enough to be important. |
| E. | No. The plot shows that stopping distances go down as temperature increases |

5. If stopping distance was expressed in yards instead of feet, how would the correlation *r* between temperatures and stopping distance change?

| | |
|---|---|
| A. | *r* would be divided by 12. |
| B. | *r* would be divided by 3. |
| C. | *r* would not change. |
| D. | *r* would be multiplied by 3. |
| E. | *r* would be multiplied by 12. |

6. If another data point were added with an air temperature of 0° C and a stopping distance of 80 feet, the correlation would

| | |
|---|---|
| A. | decrease, since this new point is an outlier that does not follow the pattern in the data. |
| B. | increase, since this new point is an outlier that does not follow the pattern in the data. |
| C. | stay nearly the same, since correlation is resistant to outliers. |
| D. | increase, since there would be more data points. |
| E. | Whether this data point causes an increase or decrease cannot be determined without recalculating the correlation. |

7. Which of the following is true of the correlation *r*?

| | |
|---|---|
| A. | It is a resistant measure of association. |
| B. | $-1 \le r \le 1$. |
| C. | If *r* is the correlation between *X* and *Y*, then -*r* is the correlation between *Y* and *X*. |
| D. | Whenever all the data lie on a perfectly straight-line, the correlation *r* will always be equal to +1.0. |
| E. | All of the above. |

Consider the following scatterplot of amounts of CO (carbon monoxide) and NOX (nitrogen oxide) in grams per mile driven in the exhausts of cars. The least-squares regression line has been drawn in the plot.

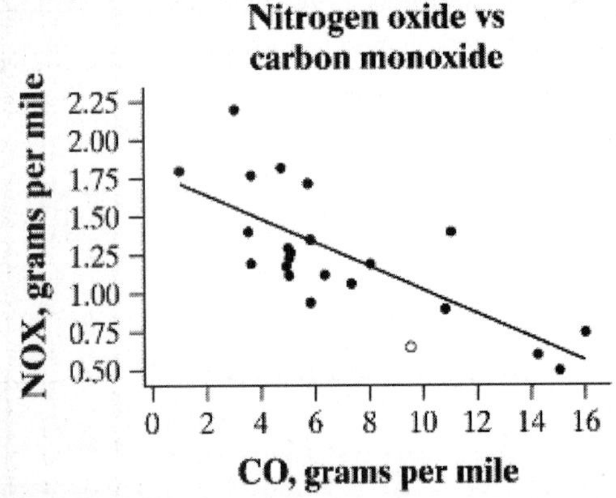


8. Based on the scatterplot, the least-squares line would predict that a car that emits 2 grams of CO per mile driven would emit approximately how many grams of NOX per mile driven?

| | |
|---|---|
| A. | 4.0 |
| B. | 1.25 |
| C. | 2.0 |
| D. | 1.7 |
| E. | 0.7 |

9. In the scatterplot, the point indicated by the open circle

| | |
|---|---|
| A. | has a negative value for the residual. |
| B. | has a positive value for the residual. |
| C. | has a zero value for the residual. |
| D. | has a zero value for the correlation. |
| E. | is an outlier. |

10. Which of the following is correct?

| A. | The correlation $r$ is the slope of the least-squares regression line. |
|---|---|
| B. | The square of the correlation is the slope of the least-squares regression line. |
| C. | The square of the correlation is the proportion of the data lying on the least-squares regression line. |
| D. | The coefficient of determination is the fraction of variability in $y$ that can be explained by least-squares regression of $y$ on $x$. |
| E. | The sum of the squared residuals from the least-squares line is 0. |

11. Which of the following statements concerning residuals from a LSRL is true?

| A. | The sum of the residuals is always 0. |
|---|---|
| B. | A plot of the residuals is useful for assessing the fit of the least-squares regression line. |
| C. | The value of a residual is the observed value of the response minus the value of the response that one would predict from the least-squares regression line. |
| D. | An influential point on a scatterplot is not necessarily the point with the largest residual. |
| E. | All of the above. |

A fisheries biologist studying whitefish in a Canadian Lake collected data on the length (in centimeters) and egg production for 25 female fish. A scatter plot of her results and computer regression analysis of egg production versus fish length are given below.
*Note that Number of eggs is given in thousands (i.e., "40" means 40,000 eggs).*

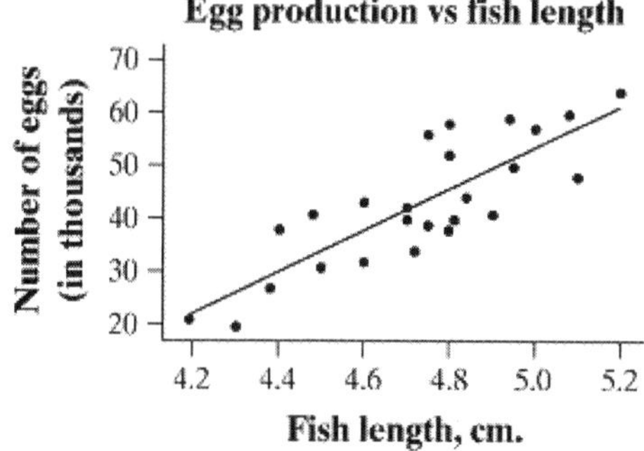


```
Predictor       Coef   SE Coef      T      P
Constant     -142.74     25.55  -5.59  0.000
Fish length   39.250     5.392   7.28  0.000

S = 6.75133    R-Sq = 69.7%    R-Sq(adj) = 68.4%
```

12. Which of the following statements is a correct interpretation of the slope of the regression line?

| A. | For each 1-cm increase in the fish length, the predicted number of eggs increases by 39.25. |
|---|---|
| B. | For each 1-cm increase in the fish length, the predicted number of eggs decreases by 142.74. |
| C. | For each 1-unit increase in the number of eggs, the predicted fish length increases by 39.25 cm. |
| D. | For each 1-unit increase in the number of eggs, the predicted fish length decreases by 142.74cm. |
| E. | For each 1-cm increase in the fish length, the predicted number of eggs increases by 39,250. |

13. What percent of variability in the number of eggs is explained by the least-squares regression of *number of eggs* on *fish length*?

| A. | 25.55 |
|---|---|
| B. | 5.392 |
| C. | 6.75133 |
| D. | 69.7 |
| E. | Cannot be determined without the original data. |

14. A study of the effects of television measured how many hours of television each of 125 grade school children watched per week during a school year and their reading scores. The study found that children who watch more television tend to have lower reading scores than children who watch fewer hours of television. The study report says that, "Hours of television watched explained 25% of the observed variation in the reading scores of the 125 subjects." The correlation between hours of TV and reading score must be

| | |
|---|---|
| A. | $r = 0.25$. |
| B. | $r = -0.25$. |
| C. | $r = -0.5$. |
| D. | $r = 0.5$. |
| E. | Can't tell from the information given. |

15. A study gathers data on the outside temperature during the winter in degrees Fahrenheit and the amount of natural gas a household consumes in cubic feet per day. Call the temperature *x* and gas consumption *y*. The house is heated with gas, so *x* helps explain *y*. The least-squares regression line for predicting *y* from *x* is: $\hat{y} = 1344 - 19x$. When the temperature goes up 1 degree, what happens to the gas usage predicted by the regression line?

| | |
|---|---|
| A. | It goes up 19 cubic feet. |
| B. | It goes down 19 cubic feet. |
| C. | It goes up 1344 cubic feet. |
| D. | It goes down 1344 cubic feet. |
| E. | Can't tell without seeing the data. |

| Problem | Answer | Concept | Right | Wrong | Simple Mistake? | Need to Study More |
|---|---|---|---|---|---|---|
| 1 | B | Explanatory vs. Response | | | | |
| 2 | A | Definition of Association | | | | |
| 3 | E | Definition of Correlation | | | | |
| 4 | C | Correlation vs. Causation | | | | |
| 5 | C | Correlation | | | | |
| 6 | A | Correlation | | | | |
| 7 | B | Correlation | | | | |
| 8 | D | Predicting with the LSRL | | | | |
| 9 | A | Residuals | | | | |
| 10 | D | Coefficient of Determination | | | | |
| 11 | E | Residuals | | | | |
| 12 | E | Slope of the LSRL | | | | |
| 13 | D | Coefficient of Determination | | | | |
| 14 | C | Coefficient of Determination | | | | |
| 15 | B | Slope of the LSRL | | | | |

## FRAPPY! Free Response AP Problem, Yay!

The following problem is modeled after actual Advanced Placement Statistics free response questions. Your task is to generate a complete, concise response in 15 minutes. After you generate your response, view two example solutions and determine whether or not you feel they are "complete," "substantial," "developing" or "minimal". If they are not "complete," what would you suggest to the student who wrote them to increase their score? Finally, you will be provided with a rubric. Score your response and note what, if anything, you would do differently to increase your own score.

A recent study was interested in determining the optimal location for fire stations in a suburban city. Ideally, fire stations should be placed so the distance between the station and residences is minimized. One component of the study examined the relationship between the amount of fire damage *y* (in thousands of dollars) and the distance between the fire station and the residence *x* (in miles). The results of the regression analysis are below.

| Predictor | Coef | SE Coef | T | P |
|---|---|---|---|---|
| Constant | 10.28 | 1.42 | 7.237 | 0.000 |
| X | 4.92 | 0.39 | 12.525 | 0.000 |

```
s = 2.232     R-Sq = 0.9235  R-Sq(adj) = 0.9176
```

(a) Write the equation of the least squares regression line. Define any variables used. Interpret the slope of the equation in context.

(b) A home located 3 miles from the fire station received $22,300 in damage. Use your equation in part (a) to calculate and interpret the residual for this observation.

(c) Identify and interpret the correlation coefficient.

**Student Response 1:**

a) $\hat{y} = 10.28 + 4.92x$
For each additional mile between the fire station and residence, we predict about $4920 additional dollars in damages.

b) $\hat{y} = 10.28 + 4.92(3) = 25.04$. Residual = 25.04 – 22.3 = 2.74. Our model overpredicted the amount of damage for this observation by $2740.

c) $r^2 = 0.9235$. There is a strong, positive linear relationship between the distance between a fire station and residence and the resulting damage in a fire.

How would you score this response? Is it substantial? Complete? Developing? Minimal? Is there anything this student could do to earn a better score?

**Student Response 2:**

a) $\widehat{firedamage} = 4.92\textit{distance} + 10.28$
We predict about $4920 additional dollars in damage for each increase of one mile between the fire station and residence that is on fire.

b) $\widehat{damage} = 2.92(3) + 10.28 = 25.04$
residual = 22.3 – 25.04 = - 2.74. Our model overpredicts the damage amount by $2740.

c) r = 0.96. There is a very strong, positive, linear relationship between a residence's damage from a fire and its distance from a fire station.

How would you score this response? Is it substantial? Complete? Developing? Minimal? Is there anything this student could do to earn a better score?

## Scoring Rubric

Use the following rubric to score your response. Each part receives a score of "Essentially Correct," "Partially Correct," or "Incorrect." When you have scored your response, reflect on your understanding of the concepts addressed in this problem. If necessary, note what you would do differently on future questions like this to increase your score.

### Intent of the Question

The goal of this question is to determine your ability to interpret computer regression output and explain key concepts of linear regression.

### Solution

**(a)** $\widehat{firedamage}$ = 10.28 + 4.92*distance* OR $\hat{y}$ = 10.28 + 4.92*x* with *x* and *y* defined as distance and damage.

For each additional mile between the fire station and residence, we predict about $4920 additional dollars in damages.

**(b)** $\widehat{damage}$ = 2.92(3) + 10.28 = 25.04
residual = 22.3 – 25.04 = - 2.74.
The model overpredicts the damage amount by $2740.

**(c)** Since $r^2$ = 0.9325, r = 0.96. There is a very strong, positive, linear relationship between a residence's damage from a fire and its distance from a fire station.

### Scoring

Parts (a), (b), and (c) are scored as essentially correct (E), partially correct (P), or incorrect (I).

**Part (a)** is essentially correct if the response (1) correctly identifies the least-squares regression equation in context or with variables defined and (2) correctly interprets the slope
Part (a) is partially correct if the response fails to define the variables in context or reverses the coefficients OR if the slope is not correctly defined in context (eg, predicts 4.92 dollars instead of $4920).

**Part (b)** is essentially correct if (1) the correct residual is calculated and (2) the interpretation is correct.
Part (b) is partially correct if only one of the above elements is correct.

**Part (c)** is essentially correct if the correlation coefficient is correctly identified and interpreted correctly with all three elements (strong, positive, linear).
Part (c) is partially correct if one of the elements (strong, positive, linear) is missing OR if $r^2$ is used instead of *r*.

4 **Complete Response**
All three parts essentially correct

3 **Substantial Response**
Two parts essentially correct and one part partially correct

2 **Developing Response**
Two parts essentially correct and no parts partially correct
One part essentially correct and two parts partially correct
Three parts partially correct

1 **Minimal Response**
One part essentially correct and one part partially correct
One part essentially correct and no parts partially correct
No parts essentially correct and two parts partially correct

# Chapter 3: Describing Relationships

**Across**

2. the difference between an observed value of the response and the value predicted by a regression line
7. Important note: Association does not imply ______.
10. graphical display of the relationship between two quantitative variables
11. line that describes the relationship between two quantitative variables
14. the coefficient of _______ describes the fraction of variability in y values that is explained by least squares regression on x.
15. A _____ association is defined when above average values of one variable are accompanied by below average values of the other.
16. individual points that substantially change the correlation or slope of the regression line

**Down**

1. the use of a regression line to make a prediction far outside the observed x values
3. the amount by which y is predicted to change when x increases by one unit
4. The ______ of a relationship in a scatterplot is determined by how closely the point follow a clear form.
5. the ____-_____ regression line is also known as the line of best fit (2 words)
6. an individual value that falls outside the overall pattern of the relationship
7. value that measures the strength of the linear relationship between two quantitative variables
8. A ______ association is defined when above average values of the explanatory are accompanied by above average values of the response
9. y-hat is the _______ value of the y-variable for a given x
11. variable that measures the outcome of a study
12. variable that may help explain or influence changes in another variable
13. The _____ of a scatterplot indicates a positive or negative association between the variables.
17. The ____ of a scatterplot is usually linear or nonlinear.

# Chapter 4: Designing Studies

*"Not everything that can be counted counts; and not everything that counts can be counted." George Gallup*

## Chapter Overview

The first three chapters introduced us to some of the basics of exploratory data analysis. In this chapter, we'll learn about the second major topic in statistics – planning and conducting a study. Since it is difficult to perform a descriptive statistical analysis without data, we need to learn appropriate ways to produce data. We will start by learning the difference between a population and a sample. Then, we will study sampling techniques and learn how to identify potential sources of bias. Our goal is to collect data that is representative of the population we wish to study. Therefore, it is important that our data collection techniques do not systematically over- or under-represent any segment of the population. In the second section, we will learn the difference between observational studies and experiments. We will learn a number of different ways to design experiments so we can establish relationships between variables. Finally, we will conclude this chapter by reviewing some cautions about using studies wisely. This chapter contains a lot of vocabulary and concepts. Be sure to study them, as proper application of terms is important for strong statistical communication!

## Sections in this Chapter



## Plan Your Learning

Use the following *suggested* guide to help plan your reading and assignments. Note: your teacher may schedule a different pacing. Be sure to follow his or her instructions!

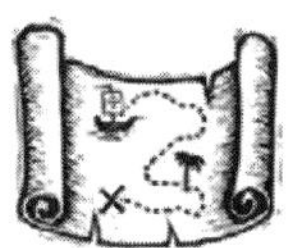

| **Read** | Intro: pp 205-207<br>4.1: pp 207-215 | 4.1: pp 215-221 | 4.1:<br>pp 221-225 | 4.2: pp 231-236 |
|---|---|---|---|---|
| **Do** | 1, 3, 5, 7, 9, 11 | 17, 19,<br>21, 23, 25 | 27-29<br>31, 33, 35 | 37-42, 45,<br>47, 49, 51, 53 |

| **Read** | 4.2:<br>pp 236-242 | 4.2:<br>pp 242-246 | 4.2: pp 246-252 | 4.3: 261-268 | Chapter Summary |
|---|---|---|---|---|---|
| **Do** | 57, 63, 65, 67 | 69, 71, 73, 75 | 77, 79, 81, 85 | 91-98,<br>102-108 | Multiple Choice<br>FRAPPY! |

## Section 4.1: Sampling and Surveys

### Before You Read: Section Summary

Often in statistics, our goal is to draw a conclusion about a population based on information gathered from a sample. In order for us to make a valid inference about the population, we must feel confident that the sample obtained will be representative of the group as a whole. There are a number of different ways to select samples from a population, some better than others. In this section, you will explore ways in which you can sample badly, how to sample well, and cautions to consider when sampling. You will be introduced to a number of different sampling methods. Be sure to familiarize yourself with how to select samples using each of the methods and how to explain potential advantages and disadvantages of each.

### "Where Am I Going?"

**Learning Targets:**

____ I can identify the population and sample in a sampling situation.

____ I can describe voluntary response and convenience samples and explain how these methods can lead to bias.

____ I can describe how to select simple random samples, stratified random samples, and cluster samples.

____ I can describe how to use a table of random digits or a random number generator to select a simple random sample.

____ I can explain how undercoverage, nonresponse, and question wording can lead to bias in a survey.

### While You Read: Key Vocabulary and Concepts

population

sample

sample survey

convenience sample

bias

voluntary response sample

simple random sample

table of random digits

stratified random sample

strata

cluster sample

inference

sampling errors

undercoverage

nonsampling errors

nonresponse

response bias

wording of questions

**After You Read: "Where Am I Now?"**
**Check for Understanding**

***Concept 1: Population, Samples, and Inference***

The purpose of sampling is to provide us with information about a population without actually gathering the information from every single element of the whole group. A sample is a part of the population from which we collect information. This information is then used to infer something about the population. We can have more faith in our inference if we are confident that the sampling method used is likely to produce a sample that is representative of the population of interest. Some sampling methods may introduce a level of bias into the situation that could cause us to make an incorrect inference about the population. It is important that you can clearly identify the population of interest and describe whether or not the sampling method is unbiased.

**Check for Understanding:** ____ *I can identify the population and sample in a sampling situation.*

A publisher is interested in determining the reading difficulty of mathematics textbooks. Reading difficulty is determined by the length of sentences and the length of words used in the text. Researchers randomly select 10 paragraphs out of the most popular Algebra 1 textbooks and calculate the average sentence length and average word length for that type of mathematic textbook. Identify the population and sample.

***Concept 2: How to Sample Well***

The key to sampling is to use a method that helps ensures that the sample is as representative of the population as possible. Because sampling methods like voluntary response and convenience samples can systematically favor certain outcomes in a population, we say they are biased. An unbiased sampling method is one that does not favor any element of the population. We rely on the use of chance to select unbiased samples. The easiest way to do this is to select a simple random sample (SRS) from the population of interest. One way we could select a SRS would be to write the name of each individual in the population on a piece of paper, put the pieces in a hat, mix them well, and draw out the necessary number of slips for our sample. We could also select a SRS by labeling each individual in the population with a number of the same length (i.e. one-, two-, three-digit number depending on the size of the population) and then generate random numbers from technology or a random number table until you reach the desired sample size. The labeled individuals that match the generated numbers, ignoring repeats, make up the sample.

**Check for Understanding:** ____ *I can describe how to use a table of random numbers or a random number generator to select a simple random sample.*

An alphabetized list of student names is found below. Use the random digit table provided to choose an SRS of 4 individuals. Clearly indicate how you are using the table to select the SRS.

| | | | | |
|---|---|---|---|---|
| Abney | Brock | Greenberg | Osters | Tyson |
| Andreasen | Bush | Knott | Preble | Wilcock |
| Bearden | Chauvet | Lacey | Ripp | Yankay |
| Bready | Costello | Martin | Rohnkol | |
| Buckley | Derksen | McDonald | Sterken | |

19223 95034 05756 28713 96409 12531 42544 82853

73676 47150 99400 11327 27754 40548 82421 76290

While Simple Random Samples give each group of $n$ individuals in the population an equal chance to be selected, they are not always the easiest to obtain. Several other sampling methods are available that can be used instead of an SRS. If the population consists of several groups of individuals that are likely to produce similar responses within groups, but systematically different responses between groups, consider taking a stratified random sample. To do this, select an SRS from each stratum (group) to obtain the sample. Another method, cluster sampling, divides the population into smaller groups (clusters) that mirror the overall population and selects an SRS of those clusters. You should be able to describe how to select a sample using each of these methods as you may be asked to do so on the AP exam!

***Concept 3: What Can Go Wrong?***

When designing a sample survey, random sampling helps avoid bias. But there are other kinds of mistakes in the sampling process that can lead to inaccurate information about the population. For instance, if the sampling method is designed in a way that leaves out certain segments of the population, the sample survey suffers from undercoverage. Bias can also result from nonsampling errors. When selected individuals cannot be contacted or refuse to participate, the survey suffers from nonresponse. If people give incorrect or misleading responses, a survey suffers from response bias. Finally, if a question is worded in a way that favors certain responses, the survey suffers from question wording bias. Be sure to design sample surveys to avoid these issues and make sure you can identify them in existing studies.

## Section 4.2: Experiments

**Before You Read: Section Summary**

Sample surveys allow us to gather information about the population without actually doing anything to that population. Such observational studies provide a snapshot of the population, but cannot be used to establish any sort of cause-effect relationship. Observational studies only allow us to describe the population, compare groups, or examine basic relationships between variables. In this section, we will move beyond sampling to study the elements of experimental design. Experiments allow us to produce data in a way that can lead to conclusions about causation. There are a lot of vocabulary terms in this section. It is easy to confuse sampling terms and experimental design terms. Make sure you understand each vocabulary term or concept!

**"Where Am I Going?"**

**Learning Targets:**

- ____ I can distinguish between an observational study and an experiment.
- ____ I can explain how lurking variables can lead to confounding.
- ____ I can identify experimental units, explanatory variables, treatments, and response variables in an experiment.
- ____ I can describe a completely randomized design for an experiment.
- ____ I can explain the meaning and purpose of blinding in an experiment.
- ____ I can distinguish between completely randomized designs and randomized block designs.
- ____ I can explain why randomize assignment is an important element of experimental design.
- ____ I can describe a randomized block design, including a matched pairs design for an experiment.

**While You Read: Key Vocabulary and Concepts**

observational study:

experiment:

lurking variable:

confounding:

experimental units:

subjects:

explanatory variables (factors):

treatments (factor levels):

completely randomized design:

control:

random assignment:

replication:

placebo:

double-blind:

statistically significant:

block:

randomized block design:

matched pairs design:

**After You Read: "Where Am I Now?"**
**Check for Understanding**

***Concept 1: Observational Studies vs. Experiments***

Sample surveys are examples of observational studies. Their goal is to describe the population and examine relationships between variables. Often, however, we wish to determine whether or not a cause-effect relationship exists between an explanatory and a response variable. Observational studies cannot be used to establish this relationship because lurking variables can result in confounding of our results.

**Check for Understanding:** ____*I can explain how lurking variables can lead to confounding*

Does shoe size affect spelling ability? A recent study was conducted in a suburban school district to answer this question. 30 students from grades 1 through 8 were randomly selected. Each student was administered a spelling test and had his or her feet measured. Test scores were plotted against shoe size and a strong, positive relationship was observed.

1. Was this an observational study or an experiment?

2. What are the explanatory and response variables?

3. Suggest a possible confounding variable in this setting. Explain carefully how it may confound the results.

***Concept 2: How to Experiment Well***

If we wish to establish cause and effect, we must conduct an experiment in which treatments are imposed on experimental units and other potential influences on the response variable are controlled as much as possible. The basic idea behind an experiment is that we obtain experimental units, apply treatments, and measure the results. When conducted properly, experiments can provide good evidence of causation. Proper experimental design incorporates three principles. First, we must control for the influences of lurking variables that might affect the response. We want to ensure that any changes are due to the treatment alone. Second, we must randomly assign experimental units to treatments. This helps ensure equivalent groups of units. Random assignment helps ensure that the effects of lurking variables will be felt equally by all groups in the experiment, so that the only systematic difference between groups is the treatments themselves. Finally, replication is necessary to ensure enough experimental units are in each group to convince us the differences in the effects of the treatments can be separated from chance differences between the groups. If a difference between groups is observed that is too large to have occurred by chance alone, we say it is statistically significant. We will learn how to establish statistical significance in a later chapter. For now, focus on how to design a quality experiment!

**Check for Understanding:** ____*I can identify experimental units, explanatory variables, treatments, and response variables in an experiment and* ____ *I can describe a completely randomized design for an experiment*

Mr. Tyson teaches statistics to 150 students. He is interested in knowing whether or not listening to classical music while studying results in higher test scores than listening to no music. He wishes to design an experiment to answer this question.

1. What are the experimental units, explanatory variable, treatments and response variable?

2. What are potential lurking variables in this situation? How could they affect the results? How could we avoid their effects?

3. Describe a completely randomized design for Mr. Tyson's experiment.

***Concept 3: Other Types of Experimental Design***

Completely randomized experiments are the simplest design that can give good evidence of cause-effect relationships. In some cases, however, we can add elements to this basic design to control for the effects of lurking variables. For example, placebos (fake treatments) can be given to control for the placebo effect—the phenomenon that occurs when subjects respond to getting any treatment, whether it is real or not. We can use blinding when there is a concern that knowing who receives what treatment might affect the results. Double-blind experiments ensure that neither the subjects nor the people who interact with the subjects and measure their responses know who receives which treatment.

When groups of subjects share a common characteristic that might systematically affect their responses to the treatments, we can use blocking to control for the effects of this lurking variable. For example, suppose researchers are conducting an experiment to compare the effectiveness of a new medicine for treating high blood pressure with the most commonly prescribed drug. If they believe that older and younger people may respond differently to such medications, the researchers can separate the subjects into blocks of older and younger people and randomly assign treatments within each block. This randomized block design helps isolate the variation in responses due to age, which makes it easier for researchers to find evidence of a treatment effect. If we are only comparing two treatments, we can sometimes conduct a matched pairs design by creating blocks of two similar individuals and then randomly assigning each subject to a treatment. Another type of matched pairs design involves assigning two treatments to each subject in a random order.

In each type of experimental design, our goal is to ensure control, random assignment, and replication. If these principles are addressed in our design, we can establish good evidence of causal relationships. Make sure you can describe how you would conduct an experiment using each of these designs as you may be asked to do so on the AP exam!

**Check for Understanding:** ____ *I can distinguish between completely randomized designs and block designs*

Refer to the previous Check for Understanding. Describe an experimental design involving blocking that will help answer Mr. Tyson's question. Explain why this design is preferable to a completely randomized design.

## Section 4.3: Using Studies Wisely

**Before You Read: Section Summary**
A common mistake on the AP exam is to confuse inference about the population and inference about cause and effect. In this section, we learn about the difference between the two as well as some of the challenges of establishing causation. While it is important to know how to design good surveys and experiments, it is just as important to understand how to properly use their results!

**"Where Am I Going?"**

**Learning Targets:**

____ I can determine the scope of inference for a statistical study.
____ I can evaluate whether a statistical study has been carried out in an ethical manner.

**While You Read: Key Vocabulary and Concepts**

inference about the population:

inference about cause and effect:

lack of realism:

*Optional (non-AP) material*
institutional review board:

informed consent:

**After You Read: "Where Am I Now?"**
**Check for Understanding**

***Concept 1: Scope of Inference***
Well-designed sample surveys allow us to make inferences about the population from which we sampled. Random sampling is what allows us to generalize our results with confidence. If the goal is to make an inference about cause and effect, we must use a randomized experiment. Unless the experimental units were randomly selected from a larger population of interest, we cannot extend our conclusions beyond individuals like those who took part in the experiment.

***Concept 2: The Challenges of Establishing Causation***
Well-designed experiments can be used to establish causation. However, lack of realism in some experiments prevents us from seeing similar results outside the laboratory setting. In some cases, it is not practical, safe, or ethical to conduct an experiment. Even with strong evidence from observational studies, it is very difficult to establish a cause and effect conclusion.

***Concept 3: Data Ethics (optional)***

Because some sample surveys and experiments have potential to cause harm to the participants, it is important to consider data ethics when designing a study. Basic data ethics include having an institutional review board that monitors the wellbeing of the participants. Individuals…consider when designing statistical studies.

## Chapter Summary: Designing Studies

This chapter is an important one in your study of statistics. After all, we cannot describe or analyze data without collecting it first! Since one of the major goals of statistics is to make inferences that go beyond the data, it is critical that we produce data in a way that will allow for such inferences. Biased data production methods can lead to incorrect inferences. Random sampling allows us to make an inference about the population as a whole. Well-designed experiments in which we randomly assign treatments and control for lurking variables allow us to make inferences about cause and effect. We will learn how to perform these inferences in later chapters. Your goal in this chapter is to be able to describe good sampling and experimental design techniques and recognize when sampling or experimental design has been done poorly. There is almost always a question about sampling or experimental design on the free-response portion of the AP exam. Be sure to familiarize yourself with all of the vocabulary and concepts from this chapter so you can answer that question with confidence!

### After You Read: "How Can I Close the Gap?"

Complete the vocabulary puzzle, multiple choice questions, and FRAPPY. Check your answers and your performance on each of the targets.

| Target | Got It! | Almost There | Needs Some Work |
|---|---|---|---|
| I can identify the population and sample in a sampling situation. | | | |
| I can describe voluntary response and convenience samples and explain how these methods can lead to bias. | | | |
| I can describe how to select simple random samples, stratified random samples, and cluster samples. | | | |
| I can describe how to use a table of random numbers or a random number generator to select a simple random sample. | | | |
| I can explain how undercoverage, nonresponse, and question wording can lead to bias in a survey. | | | |
| I can distinguish between an observational study and an experiment. | | | |
| I can explain how lurking variables can lead to confounding. | | | |
| I can identify experimental units, explanatory variables, treatments, and response variables in an experiment. | | | |
| I can describe a completely randomized design for an experiment. | | | |
| I can explain the meaning and purpose of blinding in an experiment. | | | |
| I can distinguish between completely randomized designs and randomized block designs. | | | |
| I can explain why random assignment is an important principle of experimental design. | | | |
| I can describe a randomized block design, including a matched pairs design for an experiment. | | | |
| I can determine the scope of inference for a statistical study. | | | |
| I can evaluate whether a statistical study has been carried out in an ethical manner. | | | |

Did you check "Needs Some Work" for any of the targets? If so, what will you do to address your needs for those targets?

*Learning Plan:*

## Chapter 4 Multiple Choice Practice

**Directions.** *Identify the choice that best completes the statement or answers the question. Check your answers and note your performance when you are finished.*

1. A researcher is testing a company's new stain remover. He has contracted with 40 families who have agreed to test the product. He randomly assigns 20 families to the group that will use the new stain remover and 20 to the group that will use the company's current product. The most important reason for this random assignment is that

| | |
|---|---|
| A. | randomization makes the analysis easier since the data can be collected and entered into the computer in any order. |
| B. | randomization eliminates the impact of any confounding variables. |
| C. | randomization is a good way to create two groups of 20 families that are as similar as possible, except for the treatments they receive. |
| D. | randomization ensures that the study is double-blind. |
| E. | randomization reduces the impact of outliers. |

2. A researcher observes that, on average, the number of traffic violations in cities with Major League Baseball teams is larger than in cities without Major League Baseball teams. The most plausible explanation for this observed association is that the

| | |
|---|---|
| A. | presence of a Major League Baseball team causes the number of traffic incidents to rise (perhaps due to the large number of people leaving the ballpark). |
| B. | high number of traffic incidents is responsible for the presence of Major League Baseball teams (more traffic incidents means more people have cars, making it easier for them to get to the ballpark). |
| C. | association is due to the presence of a lurking variable (Major League teams tend to be in large cities with more people, hence a greater number of traffic incidents). |
| D. | association makes no sense, since many people take public transit or walk to baseball games. |
| E. | observed association is purely coincidental. It is implausible to believe the observed association could be anything other than accidental. |

3. A researcher is testing the effect of a new fertilizer on crop growth. He marks 30 plots in a field, splits the plots in half, and randomly assigns the new fertilizer to one half of the plot and the old fertilizer to the other half. After 4 weeks, he measures the crop yield and compares the effects of the two fertilizers. This design is an example of

| | |
|---|---|
| A. | matched pairs experiment. |
| B. | completely randomized comparative experiment. |
| C. | cluster experiment. |
| D. | double-blind experiment. |
| E. | this is not an experiment. |

4. A large suburban school wants to assess student attitudes towards their mathematics textbook. The administration randomly selects 15 mathematics classes and gives the survey to every student in the class. This is an example of a

| | |
|---|---|
| A. | multistage sample. |
| B. | stratified sample. |
| C. | cluster sample. |
| D. | simple random sample. |
| E. | convenience sample. |

5. Eighty volunteers who currently use a certain brand of medication to reduce blood pressure are recruited to try a new medication. The volunteers are randomly assigned to one of two groups. One group continues to take their current medication, the other group switches to the new experimental medication. Blood pressure is measured before, during, and after the study. Which of the following best describes a conclusion that can be drawn from this study?

| | |
|---|---|
| A. | We can determine whether the new drug reduces blood pressure more than the old drug for anyone who suffers from high blood pressure. |
| B. | We can determine whether the new drug reduces blood pressure more than the old drug for individuals like the subjects in the study. |
| C. | We can determine whether the blood pressure improved more with the new drug than with the old drug, but we can't establish cause and effect. |
| D. | We cannot draw any conclusions, since the all the volunteers were already taking the old drug when the experiment started. |
| E. | We cannot draw any conclusions, because there was no control group. |

6. To determine employee satisfaction at a large company, the management selects an SRS of 200 workers from the marketing department and a separate SRS of 50 workers from the sales department. This kind of sample is called a

| | |
|---|---|
| A. | simple random sample. |
| B. | simple random sample with blocking. |
| C. | multistage random sample. |
| D. | stratified random sample. |
| E. | random cluster sample. |

7. For a certain experiment you have 8 subjects, of which 4 are female and 4 are male. The names of the subjects are listed below:

Males: Atwater, Bacon, Chu, Diaz. Females: Johnson, King, Liu, Moore

There are two treatments, A and B. If a randomized block design is used, with the subjects blocked by their gender, which of the following is <u>not</u> a possible group of subjects who receive treatment A?

| | |
|---|---|
| A. | Atwater, Chu, King, Liu |
| B. | Bacon, Chu, Liu, Moore |
| C. | Atwater, Diaz, Liu, King |
| D. | Atwater, Bacon, Chu, Johnson |
| E. | Atwater, Bacon, Johnson, King |

8. An article in the student newspaper of a large university had the headline "A's swapped for evaluations?" Results showed that higher grades directly corresponded to a more positive evaluation. Which of the following would be a valid conclusion to draw from the study?

| | |
|---|---|
| A. | A teacher can improve his or her teaching evaluations by giving good grades. |
| B. | A good teacher, as measured by teaching evaluations, helps students learn better, resulting in higher grades. |
| C. | Teachers of courses in which the mean grade is higher apparently tend to have above-average teaching evaluations. |
| D. | Teaching evaluations should be conducted before grades are awarded. |
| E. | All of the above |

9. A new cough medicine was given to a group of 25 subjects who had a cough due to the common cold. 30 minutes after taking the new medicine, 20 of the subjects reported that their coughs had disappeared. From this information you conclude

| | |
|---|---|
| A. | that the remedy is effective for the treatment of coughs. |
| B. | nothing, because the sample size is too small. |
| C. | nothing, because there is no control group for comparison. |
| D. | that the new treatment is better than the old medicine. |
| E. | that the remedy is not effective for the treatment of coughs. |

10. 100 volunteers who suffer from anxiety take part in a study. 50 are selected at random and assigned to receive a new drug that is thought to be extremely effective in reducing anxiety. The other 50 are given an existing anti-anxiety drug. A doctor evaluates anxiety levels after two months of treatment to determine if there has been a larger reduction in the anxiety levels of those who take the new drug. This would be double blind if

| | |
|---|---|
| A. | both drugs looked the same. |
| B. | neither the subjects nor the doctor knew which treatment any subject had received. |
| C. | the doctor couldn't see the subjects and the subjects couldn't see the doctor . |
| D. | there was a third group that received a placebo. |
| E. | all of the above. |

## Multiple Choice Answers

| Problem | Answer | Concept | Right | Wrong | Simple Mistake? | Need to Study More |
|---|---|---|---|---|---|---|
| 1 | C | Why We Randomize | | | | |
| 2 | C | Confounding | | | | |
| 3 | A | Matched Pairs | | | | |
| 4 | C | Cluster Sampling | | | | |
| 5 | B | Inference About the Population | | | | |
| 6 | D | Stratified Random Sampling | | | | |
| 7 | D | Blocking | | | | |
| 8 | C | Surveys vs. Experiments | | | | |
| 9 | C | Lurking Variables | | | | |
| 10 | B | Definition of Experiments | | | | |

## FRAPPY! Free Response AP Problem, Yay!

The following problem is modeled after actual Advanced Placement Statistics free response questions. Your task is to generate a complete, concise response in 15 minutes. After you generate your response, view two example solutions and determine whether you feel they are "complete," "substantial," "developing," or "minimal." If they are not "complete," what would you suggest to the student who wrote them to increase their score? Finally, you will be provided with a rubric. Score your response and note what, if anything, you would do differently to increase your own score.

A large school district is interested in determining student attitudes about their co-curricular offerings such as athletics and fine arts. The district consists of students attending 4 elementary schools (2000 students total), 1 middle school (1000 students total), and 2 high schools (2000 students total).

The administration is considering two sampling plans. The first consists of taking a simple random sample of students in the district and surveying them. The second consists of taking a stratified random sample of students and surveying them.

(a) Describe how you would select a simple random sample of 200 students in the district.

(b) Describe how you would select a stratified random sample consisting of 200 students.

(c) Describe the statistical advantage of using a stratified random sample over the simple random sample in this study.

**Student Response 1:**

a) Write the names of all 5,000 students on separate slips of paper. Place the slips into a large bin and mix them well. Draw slips of paper until you have 200.

b) Separate the students by level—elementary, middle, and high school. Label the students at each level and randomly select 66 elementary students, 66 middle school students, and 68 high school students.

c) By stratifying, we avoid surveying only elementary students or only high school students. This is important because student attitudes might be different at each level.

How would you score this response? Is it substantial? Complete? Developing? Minimal? Is there anything this student could do to earn a better score?

**Student Response 2:**

a) Label each student with a number from 0001 to 5000. Use your calculator to generate 200 random numbers. These numbers correspond to the individuals who will be surveyed.

b) Randomly select 1 elementary school, 1 middle school, and 1 high school. Randomly select 200 students from each school.

c) Stratifying is easier because we don't have to sample the entire population. It is less time consuming and gives better results.

How would you score this response? Is it substantial? Complete? Developing? Minimal? Is there anything this student could do to earn a better score?

## Scoring Rubric

Use the following rubric to score your response. Each part receives a score of "Essentially Correct," "Partially Correct," or "Incorrect." When you have scored your response, reflect on your understanding of the concepts addressed in this problem. If necessary, note what you would do differently on future questions like this to increase your score.

### Intent of the Question

The goal of this question is to determine your ability to describe sampling methods and explain the advantages of stratifying over simple random sampling

### Solution

**(a)** Write each student's name on a slip of paper. Place the slips of paper in a hat and mix well. Select 200 slips of paper and note the students in the sample. OR Label each student with a number from 0001 to 5000. Use a random number table or technology to produce random 4 digit numbers, ignoring repeats, until 200 are determined. These 200 numbers correspond to the individuals who will be surveyed.

**(b)** Because student attitudes may differ by level of school (elementary, middle, or high school), we should stratify by level. Label students at each level and randomly select 80 elementary students, 40 middle school students, and 80 high school students. This ensures each level is represented in the same proportion as the overall student enrollments

**(c)** Stratifying ensures no level is over or under represented in the sample. It is possible to select very few (or even no!) students from one level in a simple random sample. The opinions of students at one level may not reflect the opinions of all students in the district. Stratifying ensures each level is fairly represented.

### Scoring:

Parts (a), (b), and (c) are scored as essentially correct (E), partially correct (P), or incorrect (I).

**Part (a)** is essentially correct if the response describes an appropriate method of selecting a simple random sample. This method should include labeling the individuals and employing a sufficient means of random selection that could be replicated by someone knowledgeable in statistics.
Part (a) is partially correct if random selection is used correctly, but the description does not provide sufficient detail for implementation.

**Part (b)** is essentially correct if the response describes selecting strata based on a reasonable variable (such as school level) and indicates randomly selecting individuals from each stratum to be a part of the survey. The method can result in an equal number of students from each level OR proportional representation based on the strata.

Part (b) is partially correct if a reasonable variable is identified, but the method is unclear or does not ensure proportional representation.

**Part (c)** is essentially correct if the response provides a reasonable statistical advantage of stratified random sampling based on the effects of an identified variable on the results in the context of the problem.
Part (c) is partially correct if the response provides a reasonable statistical advantage, but the communication is not clear or lacks context.

**4 Complete Response**
All three parts essentially correct

**3 Substantial Response**
Two parts essentially correct and one part partially correct

**2 Developing Response**
Two parts essentially correct and no parts partially correct
One part essentially correct and two parts partially correct
Three parts partially correct

**1 Minimal Response**
One part essentially correct and one part partially correct
One part essentially correct and no parts partially correct
No parts essentially correct and two parts partially correct

# Chapter 4: Designing Studies

### oss

a ______ random sample consists of separate simple random samples drawn from groups of similar individuals
a "fake" treatment that is sometimes used in experiments
he effort to minimize variability in the way experimental units are obtained and treated
he process of drawing a conclusion about the population based on a sample
his type of student can not be used to establish cause-effect relationships
he practice of using enough subjects in an experiment to reduce chance variation
a study that systematically favors certain outcomes shows this
his occurs when some groups in the population are left out of the process of choosing the sample
a study in which a treatment is imposed in order to observe a reasponse
he entire group of individuals about which we want information
a simple _____ sample consists of individuals from the population, each of which has an equally likely chance of being chosen
a _____ sample consists of a simple random sample of small groups from a population

### Down

1. groups of similar individuals in a population
2. a group of experimental units that are similar in some way that may affect the response to the treatments
3. the rule used to assign experimental units to treatments is ____ assignment
5. smaller groups of individuals who mirror the population
6. this occurs when an individual chosen for the sample can't be contacted or refuses to participate
7. an observed effect that is too large to have occurred by chance alone
9. a lack of ____ in an experiment can prevent us from generalizing the results
10. a sample in which we choose individuals who are easiest to reach
12. a ____ response sample consists of people who choose themselves by responding to a general appeal
13. neither the subjects nor those measuring the response know which treatment a subject received (two words)
14. when units are humans, they are called
17. the part of the population from which we actually collect information
18. another name for treatments
19. the individuals on which an experiement is done are experimental ____

# Chapter 5: Probability – What are the Chances?

*"The most important questions of life are, for the most part, really questions of probability." Pierre-Simon LaPlace*

## Chapter Overview

Now that we have learned how to collect data and how to analyze it graphically and numerically, we turn our study to probability, the mathematics of chance. Probability is the basis for the fourth and final theme in AP Statistics, inference. The next three chapters will provide you with the background in probability necessary to perform and understand the inferential methods we'll learn later in the course.

In this chapter, you will learn the definition of probability as a long-term relative frequency. You will study how to use simulation to answer probability questions as well as some basic rules to calculate probabilities of events. You will also learn two concepts that will reappear later in our studies: conditional probability and independence. Many of the ideas and methods you will learn in this chapter may be familiar to you. When it comes to statistics, your goal with probability is to be able to answer the question, "What would happen if we did this many times?" so you can make an informed statistical inference.

## Sections in this Chapter

**Section 5.1**: Randomness, Probability, and Simulation
**Section 5.2**: Probability Rules
**Section 5.3**: Conditional Probability and Independence

## Plan Your Learning

Use the following *suggested* guide to help plan your reading and assignments. Note: your teacher may schedule a different pacing. Be sure to follow his or her instructions!

| **Read** | 5.1: pp 281-288 | 5.1: pp 289-293 | 5.2: pp 299-302 | 5.2: pp 303-308 |
|---|---|---|---|---|
| **Do** | 1, 3, 5, 7, 9, 11 | 15, 17, 19, 23, 25 | 27, 31, 32, 43, 45, 47 | 29, 33-36, 49, 51, 53, 55 |

| **Read** | 5.3: pp 312-320 | 5.3: pp 321-328 | Chapter Summary |
|---|---|---|---|
| **Do** | 57-60, 63, 65, 67, 69, 79, 77, 79 | 83, 58, 87, 91, 93, 95, 97, 99 | Multiple Choice FRAPPY! |

## Section 5.1: Randomness, Probability, and Simulation

### Before You Read: Section Summary

This section introduces the basic definition of probability as a long-term relative frequency. That is, probability answers the question, "How often would we expect to see a particular outcome if we repeated a chance process many times?" There are a lot of common misconceptions about probability. Several are discussed in this section. Be sure to avoid falling for these common myths! The last topic in this section addresses the use of simulation to estimate probabilities. Simulation is a powerful tool for modeling chance behavior that can be used to illustrate many of the inference ideas you'll study later in the course.

### "Where Am I Going?"

**Learning Targets:**

____ I can interpret probability as a long-run relative frequency
____ I can use simulation to model chance behavior

### While You Read: Key Vocabulary and Concepts

law of large numbers:

probability:

simulation:

### After You Read: "Where Am I Now?" Check for Understanding

#### *Concept 1: The Idea of Probability*

When we observe chance behavior over a long series of repetitions, a useful fact emerges. While chance behavior is unpredictable in the short term, a regular and predictable pattern becomes evident in the long run. The law of large numbers tells us that as we observe more and more repetitions of a chance behavior, the proportion of times a specific outcome occurs will "settle down" around a single value. This long-term proportion is the probability of the outcome occurring. The probability of an event is always described as a value between 0 and 1, with 0 representing it is impossible for the event to occur and 1 representing it is guaranteed the event will occur.

**Check for Understanding:** ____ *I can interpret probability as a long-run relative frequency.*

The probability of drawing a jack, queen, or king from a standard deck of playing cards is approximately 0.23.

a) Explain what this probability means in the context of drawing from a deck of cards.

b) Does this mean if we repeatedly draw a card, replace it, shuffle, and draw again 100 times that we will draw a jack, queen, or king 23 times? Why or why not?

### *Concept 2: Simulation*

We first saw an example of simulation in Chapter 1. In this section, we learn how we can use simulation to estimate the probability of an event occurring. The four-step process can be used to perform a simulation by identifying the question of interest about a chance process, describing how to use a chance device to imitate a repetition of the chance behavior, performing many repetitions of the simulation, and using the results of the simulation to answer the original question. While simulations don't provide exact theoretical probabilities, the use of random numbers and other chance devices to imitate chance behavior can be a useful tool for estimating the likelihood of events.

**Check for Understanding:** ____ *I can use simulation to model chance behavior.*

A popular airline knows that, in general, 95% of individuals who purchase a ticket for a 10-seat commuter flight actually show up for the flight. In an effort to ensure a full flight, the airline sells 12 tickets for each flight. Design and carry out a simulation to estimate the probability that the flight will be overbooked, that is, more passengers show up than there are seats on the flight.

## Section 5.2: Probability Rules

### Before You Read: Section Summary

Now that you have the basic idea of probability down, you will learn how to describe probability models and use probability rules to calculate the likelihood of events. You will also learn how to organize information in two-way tables and Venn diagrams to help in determining probabilities. Understanding probability is important for understanding inference. Make sure you are comfortable with the definitions and rules in this section as it will make your study of probability much easier!

### "Where Am I Going?"

**Learning Targets:**

_____ I can describe a probability model for a chance process

_____ I can use basic probability rules such as the complement rule and addition rule for mutually exclusive events

_____ I can find the probability that an event occurs using a two-way table

_____ I can use a Venn diagram to model a chance process involving two events

_____ I can use the general addition rule to calculate $P(A \cup B)$

### While You Read: Key Vocabulary and Concepts

sample space *S*:

probability model:

event:

complement:

mutually exclusive (disjoint):

general addition rule:

intersection:

union:

**After You Read: "Where Am I Now?"**
**Check for Understanding**

### *Concept 1: Probability Models and the Basic Rules of Probability*

Chance behavior can be described using a probability model. This model provides two pieces of information: a list of possible outcomes (sample space) and the likelihood of each outcome. By describing chance behavior with a probability model, we can find the probability of an event—a particular outcome or collection of outcomes. Probability models must obey some basic rules of probability:

- For any event *A*, $0 \le P(A) \le 1$.
- If *S* is the sample space in a probability model, $P(S) = 1$.
- In the case of equally likely outcomes, $P(A)$ = (# outcomes in event *A*) / (# outcomes in *S*).
- $P(A^C) = 1 - P(A)$.
- If *A* and *B* are mutually exclusive events, $P(A \text{ or } B) = P(A) + P(B)$.

After this section, you should be able to describe a probability model for chance behavior and apply the basic probability rules to answer questions about events.

**Check for Understanding:** ____ *I can describe a probability model for a chance process and* ____ *I can use basic probability rules such as the complement rule and addition rule for mutually exclusive events*

Consider drawing a card from a shuffled fair deck of 52 playing cards.

1. How many possible outcomes are in the sample space for this chance process? What's the probability for each outcome?

Define the following events:
*A*: the card drawn is an Ace
*B:* the card drawn is a heart

2. Find P(A) and P(B).

3. What is P(AC)?

3. Are events *A* and *B* mutually exclusive? Why or why not?

### *Concept 2: Two-Way Tables and Venn Diagrams*

Often we'll need to find probabilities involving two events. In these cases, it may be helpful to organize and display the sample space using a two-way table or Venn diagram. This can be especially helpful when two events are not mutually exclusive. When dealing with two events *A* and *B*, it is important to be able to describe the union (or collection of all outcomes in *A*, *B*, or both) and the intersection (the collection of outcomes in both *A* and *B*). The general addition rule expands upon the basic rules presented in this section to help us find the probability of two events that are not mutually exclusive.

- If *A* and *B* are two events, $P(A \cup B) = P(A) + P(B) - P(A \cap B)$.

**Check for Understanding:** ____ *I can use a Venn diagram to model a chance process involving two events,* ____ *I can find the probability of an event using a two-way table, and* ____ *I can use the general addition rule to calculate P(A U B)*

Consider drawing a card from a shuffled fair deck of 52 playing cards.
Define the following events:
*A*: the card drawn is an Ace
*B:* the card drawn is a heart

1. Use a two way table to display the sample space.

2. Use a Venn diagram to display the sample space.

3. Find $P(A \cup B)$. Show your work.

## Section 5.3: Conditional Probability and Independence

### Before You Read: Section Summary

Two important concepts are introduced in this section: conditional probability and independence. These concepts will reappear throughout the remainder of your studies in statistics, so it is important that you understand what they mean! This section will also introduce you to several rules for calculating probabilities: the general multiplication rule, the multiplication rule for independent events, and the conditional probability formula. Not only do you want to know how to use these rules, but also when. As you perform probability calculations, make sure you can justify why you are using a particular rule!

### "Where Am I Going?"

**Learning Targets:**

- ____ I can use a tree diagram to describe chance behavior
- ____ I can use the general multiplication rule to solve probability questions
- ____ I can compute conditional probabilities
- ____ I can determine whether two events are independent
- ____ I can use the multiplication rule for independent events to compute probabilities

### While You Read: Key Vocabulary and Concepts

conditional probability:

independent:

tree diagram:

general multiplication rule:

multiplication rule for independent events:

conditional probability formula:

### After You Read: "Where Am I Now?" Check for Understanding

#### *Concept 1: Conditional Probability and Independence*

A conditional probability describes the chance that an event will occur given that another event is already known to have happened. To note that we are dealing with a conditional probability, we use the symbol | to mean "given that." For example, suppose we draw one card from a shuffled deck of 52 playing cards. We could write "the probability that the card is an ace given that it is a red card as P(ace | red). Building on the concept of conditional probabilities, we can say that when knowing that one event has occurred has no effect on the probability of another event occurring, the events are independent. That is, events A and B are independent if P(A | B) = P(A) and P(B | A) = P(B). Note that the events "get an ace" and "get a red card" described earlier are independent.

**Check for Understanding:** ____ *I can compute conditional probabilities and* ____ *I can determine whether two events are independent*

Is there a relationship between gender and candy preference? Suppose 200 high school students were asked to complete a survey about their favorite candies. The table below shows the gender of each student and their favorite candy.

| | **Male** | **Female** | **Total** |
|---|---|---|---|
| **Skittles** | 80 | 60 | 140 |
| **M & M's** | 40 | 20 | 60 |
| **Total** | 120 | 80 | 200 |

Define $A$ to be the event that a randomly selected student is *male* and $B$ to be the event that a randomly selected student likes *Skittles*. Are the events $A$ and $B$ independent? Justify your answer.

### *Concept 2: Tree Diagrams and the Multiplication Rule*

When chance behavior involves a sequence of events, we can model it using a tree diagram. A tree diagram provides a branch for each outcome of an event along with the associated probabilities of those outcomes. Successive branches represent particular sequences of outcomes. To find the probability of an event, we multiply the probabilities on the branches that make up the event. This leads us to the general multiplication rule: $P(A \cap B) = P(A) \cdot P(B \mid A)$. If $A$ and $B$ are independent, the probability that both events occur is $P(A \cap B) = P(A) \cdot P(B)$.

**Check for Understanding:** ____ *I can use a tree diagram to describe chance behavior and* ____ *I can use the general multiplication rule to solve probability questions*

A study of high school juniors in three districts – Lakeville, Sheboygan, and Omaha – was conducted to determine enrollment trends in AP mathematics courses—Calculus or Statistics. 42% of students in the study came from Lakeville, 37% came from Sheboygan, and the rest came from Omaha. In Lakeville, 64% of juniors took Statistics and the rest took Calculus. 58% of juniors in Sheboygan and 49% of juniors in Omaha took Statistics while the rest took Calculus in each district. No juniors took both Statistics and Calculus. Describe this situation using a tree diagram and find the probability that a randomly selected student from in the study took Statistics.

### Concept 3: Calculating Conditional Probabilities

By rearranging the terms in the general multiplication rule, we can determine a rule for conditional probabilities. That is, $P(B \mid A) = P(A \cap B) / P(A)$. Most conditional probabilities can be determined by using a two-way table, Venn diagram, or tree diagram. However, the formula can also be used if you know the appropriate probabilities in the situation.

**Check for Understanding:** ____ *I can compute conditional probabilities*

Consider the situation from Concept 2. Find *P*(student is from Lakeville | took Statistics).

## Chapter Summary: Probability – What are the Chances?

Probability describes the long-term behavior of chance processes. Since chance occurrences display patterns of regularity after many repetitions, we can use the rules of probability to determine the likelihood of observing particular results. At this point, you should be comfortable with the basic definition and rules of probability. In the next two chapters, you will study some further concepts in probability so we can build the foundation necessary for statistical inference.

Note that the AP exam may contain several questions about the probability of particular events. Make sure you understand how and when to apply each formula. More importantly, make sure you show your work when calculating probabilities so anyone reading your response understands exactly how you arrived at your answer!

### After You Read: "How Can I Close the Gap?"

Complete the vocabulary puzzle, multiple choice questions, and FRAPPY. Check your answers and on your performance on each of the targets.

| Target | Got It! | Almost There | Needs Some Work |
|---|---|---|---|
| I can interpret probability as a long-run relative frequency. | | | |
| I can use simulation to model chance behavior. | | | |
| I can describe a probability model for a chance process. | | | |
| I can use basic probability rules such as the complement rule and addition rule for mutually exclusive events. | | | |
| I can use a Venn diagram to model a chance process involving two events. | | | |
| I can use the general addition rule to calculate $P(A \cup B)$. | | | |
| I can use a tree diagram to describe chance behavior. | | | |
| I can use the general multiplication rule to solve probability questions. | | | |
| I can compute conditional probabilities. | | | |
| I can determine whether two events are independent. | | | |
| I can find the probability that an event occurs using a two-way table. | | | |
| I can use the multiplication rule for independent events to compute probabilities. | | | |

Did you check "Needs Some Work" for any of the targets? If so, what will you do to address your needs for those targets?

*Learning Plan:*

# Chapter 5 Multiple Choice Practice

**Directions.** *Identify the choice that best completes the statement or answers the question. Check your answers and note your performance when you are finished.*

1. The probability that you will win a prize in a carnival game is about 1/7. During the last nine attempts, you have failed to win. You decide to give it one last shot. Assuming the outcomes are independent from game to game, the probability that you will win is:

| | |
|---|---|
| A. | 1/7 |
| B. | (1/7) - (1/7)^9 |
| C. | (1/7) + (1/7)^9 |
| D. | 1/10 |
| E. | 7/10 |

2. A friend has placed a large number of plastic disks in a hat and invited you to select one at random. He informs you that half are red and half are blue. If you draw a disk, record the color, replace it, and repeat 100 times, which of the following is true?

| | |
|---|---|
| A. | It is unlikely you will choose red more than 50 times. |
| B. | If you draw 10 blue disks in a row, it is more likely you will draw a red on the next try. |
| C. | The overall proportion of red disks drawn should be close to 0.50. |
| D. | The chance that the $100^{th}$ draw will be red depends on the results of the first 99 draws. |
| E. | All of the above are true. |

3. The two-way table below gives information on males and females at a high school and their preferred music format.

| | CD | mp3 | Vinyl | Totals |
|---|---|---|---|---|
| Males | 146 | 106 | 48 | 300 |
| Females | 146 | 64 | 40 | 250 |
| Totals | 292 | 170 | 88 | 550 |

You select one student from this group at random. Which of the following statement is true about the events "prefers vinyl" and "Male"?

| | |
|---|---|
| A. | The events are mutually exclusive and independent. |
| B. | The events are not mutually exclusive but they are independent. |
| C. | The events are mutually exclusive, but they are not independent. |
| D. | The events are not mutually exclusive, nor are they independent. |
| E. | The events are independent, but we do not have enough information to determine if they are mutually exclusive. |

4. People with type O-negative blood are universal donors. That is, any patient can receive a transfusion of O-negative blood. Only 7.2% of the American population has O-negative blood. If 10 people appear at random to give blood, what is the probability that at least 1 of them is a universal donor?

| | |
|---|---|
| A. | 0 |
| B. | 0.280 |
| C. | 0.526 |
| D. | 0.720 |
| E. | 1 |

5. A die is loaded so that the number 6 comes up three times as often as any other number. What is the probability of rolling a 4, 5, or 6?

| | |
|---|---|
| A. | 2/3 |
| B. | 1/2 |
| C. | 5/8 |
| D. | 1/3 |
| E. | 1/4 |

6. You draw two candies at random from a bag that has 20 red, 10 green, 15 orange, and 5 blue candies without replacement. What is the probability that both candies are red?

| | |
|---|---|
| A. | 0.1551 |
| B. | 0.1600 |
| C. | 0.2222 |
| D. | 0.4444 |
| E. | 0.8000 |

7. An event A will occur with probability 0.5. An event B will occur with probability 0.6. The probability that both A and B will occur is 0.1.

| | |
|---|---|
| A. | Events A and B are independent. |
| B. | Events A and B are mutually exclusive. |
| C. | Either A or B always occurs. |
| D. | Events A and B are complementary. |
| E. | None of the above is correct. |

8. Event A occurs with probability 0.8. The conditional probability that event B occurs, given that A occurs, is 0.5. The probability that both A and B occur is:

| | |
|---|---|
| A. | 0.3 |
| B. | 0.4 |
| C. | 0.625 |
| D. | 0.8 |
| E. | 1.0 |

9. At Lakeville South High School, 60% of students have high-speed internet access, 30% have a mobile computing device, and 20% have both. The proportion of students that have neither high-speed internet access nor a mobile computing device is:

| | |
|---|---|
| A. | 0% |
| B. | 10% |
| C. | 30% |
| D. | 80% |
| E. | 90% |

10. Experience has shown that a certain lie detector will show a positive reading (indicates a lie) 10% of the time when a person is telling the truth and 95% of the time when a person is lying. Suppose that a random sample of 5 suspects is subjected to a lie detector test regarding a recent one-person crime. The probability of observing no positive readings if all suspects plead innocent and are telling the truth is:

| | |
|---|---|
| A. | 0.409 |
| B. | 0.735 |
| C. | 0.00001 |
| D. | 0.591 |
| E. | 0.99999 |

## Multiple Choice Answers

| Problem | Answer | Concept | Right | Wrong | Simple Mistake? | Need to Study More |
|---|---|---|---|---|---|---|
| 1 | A | Probability Basics | | | | |
| 2 | C | Definition of Probability | | | | |
| 3 | B | Mutually Exclusive/Independent | | | | |
| 4 | C | Probability Calculations | | | | |
| 5 | C | Probability Calculations | | | | |
| 6 | A | Probability Calculations | | | | |
| 7 | C | Probability Basics | | | | |
| 8 | B | Conditional Probabilities | | | | |
| 9 | C | General Addition Rule | | | | |
| 10 | D | Conditional Probabilities | | | | |

## FRAPPY! Free Response AP Problem, Yay!

The following problem is modeled after actual Advanced Placement Statistics free response questions. Your task is to generate a complete, concise response in 15 minutes. After you generate your response, view two example solutions and determine whether you feel they are "complete," "substantial," "developing" or "minimal." If they are not "complete," what would you suggest to the student who wrote them to increase their score? Finally, you will be provided with a rubric. Score your response and note what, if anything, you would do differently to increase your own score.

A simple random sample of adults in a metropolitan area was selected and a survey was administered to determine political views. The results are recorded below:

| | Political Views | | | |
|---|---|---|---|---|
| Age | Conservative | Moderate | Liberal | Total |
| 18-29 | 10 | 15 | 30 | 55 |
| 30-44 | 20 | 30 | 35 | 85 |
| 45-59 | 35 | 15 | 20 | 70 |
| Over 60 | 20 | 15 | 10 | 45 |
| Total | 85 | 75 | 95 | 255 |

(a) What is the probability that a person chosen at random from this sample will have moderate political views?

(b) What is the probability that a person chosen at random from those in the sample who are between the ages of 30 and 44 will have moderate political views? Show your work.

(c) Based on your answers to (a) and (b), are political views and age independent for the population of adults in this metropolitan area? Why or why not?

**Student Response 1:**

a) 75/255

b) P(moderate | 30-44) = 30/75 = 0.40

c) Yes, the two are independent because political views don't depend on age. There are moderates in every age category.

How would you score this response? Is it substantial? Complete? Developing? Minimal? Is there anything this student could do to earn a better score?

**Student Response 2:**

a) P(moderate) = 0.29

b) P(moderate | 30-44) = P(moderate and 30-44) / P(30-44) = 30/85 = 0.35

c) No, these are not independent because P(moderate) $\neq$ P(moderate | 30-44). In order to be independent, these probabilities should be the same and the condition of age should not affect the probability of political views.

How would you score this response? Is it substantial? Complete? Developing? Minimal? Is there anything this student could do to earn a better score?

## Scoring Rubric

Use the following rubric to score your response. Each part receives a score of "Essentially Correct," "Partially Correct," or "Incorrect." When you have scored your response, reflect on your understanding of the concepts addressed in this problem. If necessary, note what you would do differently on future questions like this to increase your score.

### Intent of the Question

The goal of this question is to determine your ability to calculate probabilities and determine whether or not two events are independent.

### Solution

**(a)** *P(*moderate*)* = 75/255 = 0.2941

**(b)** *P(*moderate | *age 30-44)* = 30/85 = 0.3529

**(c)** If moderate political views and age were independent, the probabilities in (a) and (b) would be the same. Since they are not equal, age and political views are not independent for the individuals in this sample.

### Scoring:

Parts (a), (b), and (c) are scored as essentially correct (E), partially correct (P), or incorrect (I).

**Part (a)** is essentially correct if the probability is correct. Part (a) is partially correct if the correct formula is shown, but minor arithmetic errors are present. Otherwise it is incorrect.

**Part (b)** is essentially correct if the conditional probability is calculated correctly. Part (b) is partially correct if the conditioning is reversed and *P(age 30-44* | moderate*) = 30/75 = 0.40* is calculated.

**Part (c)** is essentially correct if the response indicates the two variables are not independent and justifies the conclusion based on an appropriate probability argument. Part (c) is partially correct if the response indicates the two variables are not independent, but the argument is weak or not based on an appropriate probability argument.

**4 Complete Response**
All three parts essentially correct

**3 Substantial Response**
Two parts essentially correct and one part partially correct

**2 Developing Response**
Two parts essentially correct and no parts partially correct
One part essentially correct and two parts partially correct
Three parts partially correct

**1 Minimal Response**
One part essentially correct and one part partially correct
One part essentially correct and no parts partially correct
No parts essentially correct and two parts partially correct

# Chapter 5: Probability

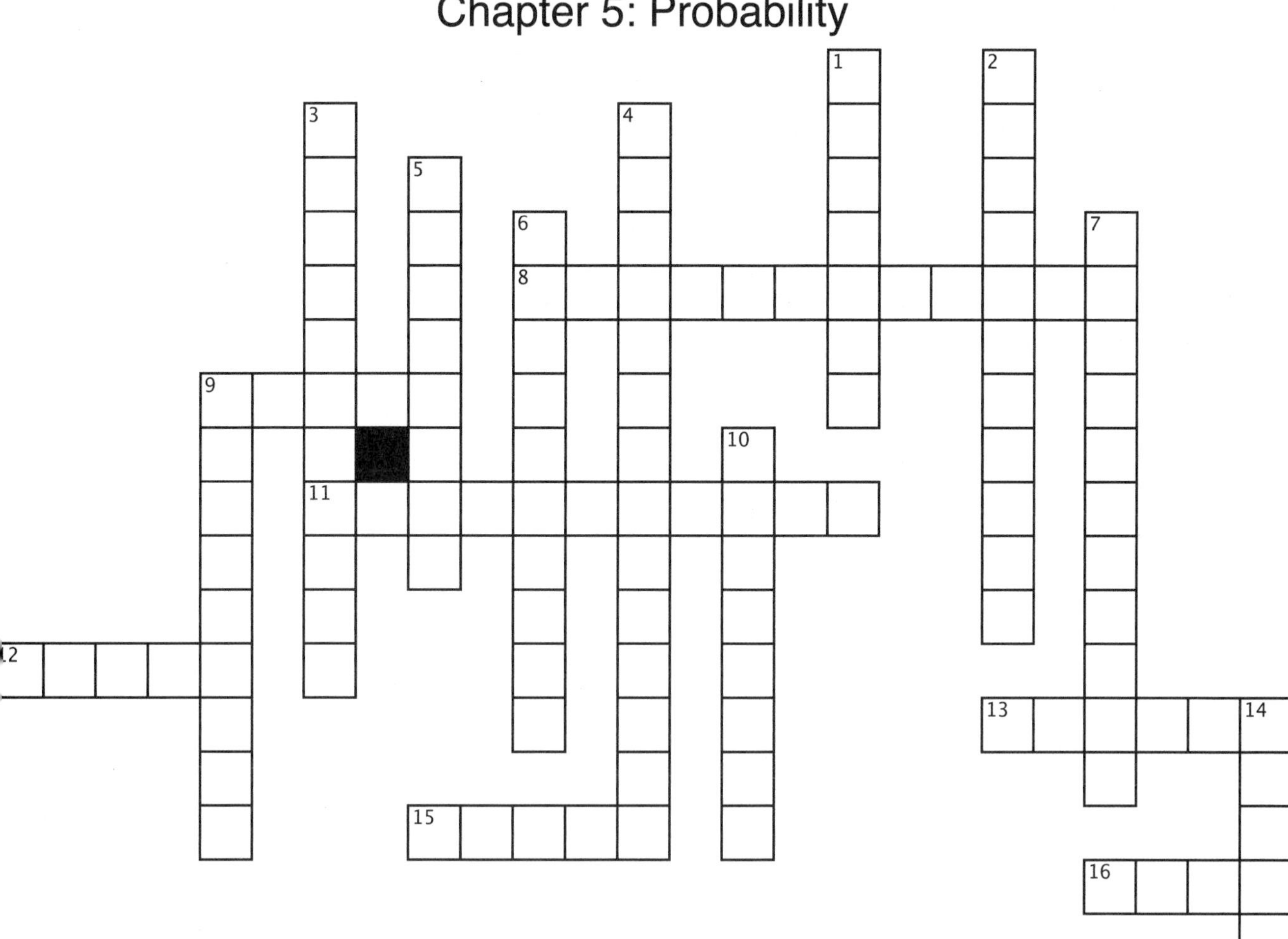

## ross

- The collection of outcomes that occur in both of two events.
- A collection of outcomes from a chance process.
- The proportion of times an outcome would occur in a very long series of repetitions.
- _____ Theorem can be used to find probabilities that require going "backward" in a tree diagram.
- In statistics, this doesn't mean "haphazard." It means "by chance."
- The collection of outcomes that occur in either of two events.
- A _____ diagram can help model chance behavior that involves a sequence of outcomes.

## Down

1. The law of large _____ states that the proportion of times an outcome occurs in many repetitions will approach a single value.
2. The probability that one event happens given another event is known to have happened.
3. The set of all possible outcomes for a chance process (two words).
4. The probability that two events both occur can be found using the general _____ rule.
5. P(A or B) can be found using the general _____ rule.
6. The imitation of chance behavior, based on a model that reflects the situation.
7. The occurrence of one event has no effect on the chance that another event will happen.
9. Another term for disjoint: Mutually _____.
10. Two events that have no outcomes in common and can never occur together.
14. A probability _____ describes a chance process and consists of two parts.

# Chapter 6: Random Variables

*"Chance, too, which seems to rush along with slack reins, is bridled and governed by law." Boethius*

## Chapter Overview

In the last chapter we learned the basic definition and rules of probability. We continue our study of probability in this chapter by exploring situations that involve assigning a numerical value to each possible outcome of a chance process. The random variables that result form the foundation for inference procedures in later chapters. You will learn how to calculate probabilities of events involving random variables as well as how to describe their probability distributions. Specifically, you will learn formulas to determine the mean and standard deviation of individual random variables as well as the combination of several independent random variables. Finally, you'll explore two special random variables – binomial and geometric – and learn how to calculate probabilities of events in binomial and geometric settings. This chapter involves a lot of formulas, so you may want to familiarize yourself with the formula sheet provided on the AP exam. Like earlier chapters, you should focus less on memorizing formulas or calculator keystrokes and more on how to apply the formulas and interpret results.

## Sections in this Chapter

**Section 6.1**: Discrete and Continuous Random Variables
**Section 6.2**: Transforming and Combining Random Variables
**Section 6.3**: Binomial and Geometric Random Variables

## Plan Your Learning

Use the following *suggested* guide to help plan your reading and assignments. Note: your teacher may schedule a different pacing. Be sure to follow his or her instructions!

| **Read** | 6.1: pp 339-346 | 6.1: pp 346-353 | 6.2: pp 358-363 | 6.2: pp 364-377 |
|---|---|---|---|---|
| **Do** | 1, 5, 7, 9, 13 | 14, 18, 19, 23, 25 | 27-30, 37, 39-41, 43, 45 | 49, 51, 57-59, 63 |

| **Read** | 6.3: pp 382-389 | 6.3: pp 390-397 | 6.3: pp 397-403 | Chapter Summary |
|---|---|---|---|---|
| **Do** | 61, 65, 66, 69, 71, 73, 75, 77 | 79, 81, 83, 85, 87, 89 | 93, 95, 97, 99, 101-103 | Multiple Choice FRAPPY! |

## Section 6.1: Discrete and Continuous Random Variables

### Before You Read: Section Summary

A random variable takes numerical values that describe the outcomes of a chance process. In the last chapter, we learned that a probability model describes the possible outcomes for a chance process and their probabilities. A random variable does the same thing, describing the possible values that the variable takes and the probability of each. Random variables fall into two categories: discrete and continuous. What differentiates the two is the set of values the random variable can take. If the set is limited to fixed values with gaps between, it is discrete. If the variable can take on any value in an interval, it is continuous. Regardless of the type, we are interested in describing the shape of the random variable's probability distribution, its center, and its spread. Knowing these characteristics will give us a sense of what to expect in repeated observations of the random variable as well as what can be considered likely and unlikely results. This idea forms the basis for inferential thinking in later chapters so you want to get used to thinking along those lines in this section!

### "Where Am I Going?"

**Learning Targets:**

- ____ I can use a probability distribution to answer questions about possible values of a random variable.
- ____ I can calculate and interpret the mean of a random variable.
- ____ I can calculate and interpret the standard deviation of a random variable.

### While You Read: Key Vocabulary and Concepts

random variable:

probability distribution:

discrete random variable:

mean (expected value) of a discrete random variable:

variance of a discrete random variable:

standard deviation of a discrete random variable:

continuous random variable:

**After You Read: "Where Am I Now?"**
**Check for Understanding**

### *Concept 1: Discrete and Continuous Random Variables*

A random variable can be classified as either discrete or continuous depending on its possible values. If it takes a fixed (finite or infinite) set of possible values with gaps in between, then we call it a discrete random variable. If the random variable takes on *any* value in an interval of numbers, it is continuous. To describe a random variable, we follow the same process as describing a probability model. First, define the random variable, X, as a numerical outcome of a chance process. Next, indicate the possible values of the variable. Finally, give the probability that each value occurs using a table, formula, or graph.

**Check for Understanding:** ____ *I can use a probability distribution to answer questions about possible values of a random variable.*

Consider two 4-sided dice, each having sides labeled 1, 2, 3, 4. Let X = the sum of the numbers that appear after a roll of the dice.

a) Is X a discrete or a continuous random variable? Sketch the probability distribution of X below. Describe what you see.

b) If somebody rolled the dice 10 times and got a sum less than 3 each time, would you be surprised? Why or why not?

### *Concept 2: Mean and Standard Deviation of Random Variables*

In Chapter 1, we learned that when describing distributions of quantitative data, we should always note the shape, center, and spread. The same holds true for random variables. In order to make inferences, we need to know what is considered a "typical" value of the random variable being examined as well as how much variation around that value we can expect to see. As with distributions of quantitative data, the center of a random variable's probability distribution can be described by calculating the mean. However, in the case of random variables, the mean (or expected value) is a *weighted* average, taking into account the probability of each outcome occurring. Likewise, the standard deviation of a random variable takes into account the probability of each outcome occurring, giving more weight to those outcomes that are more likely. Make sure you get comfortable with the formulas for the mean and standard deviation of a discrete random variable so you can calculate and interpret the center and spread of the probability distribution.

**Check for Understanding:** ____ *I can calculate and interpret the mean of a random variable* and ____ *I can calculate and interpret the standard deviation of a random variable.*

a) Suppose the random variable Y = *number of goals in a randomly selected high school hockey game* has the following probability distribution:

| Goals: | 0 | 1 | 2 | 3 | 4 |
|---|---|---|---|---|---|
| Probability: | 0.155 | 0.195 | 0.243 | 0.233 | 0.174 |

Sketch the probability distribution. Then calculate the mean and standard deviation of Y and interpret them in the context of the situation.

b) The weights of toddler boys follow an approximately Normal distribution with mean 34 pounds and standard deviation 3.5 pounds. Suppose you randomly choose one toddler boy and record his weight. What is the probability that the randomly selected boy weighs less than 31 pounds?

## Section 6.2: Transforming and Combining Random Variables

### Before You Read: Section Summary

This section introduces two distinct topics. First, you will explore transforming a single random variable. That is, you will learn how to describe the shape, center, and spread of the probability distribution of a random variable when a linear transformation (such as adding a constant to each value or multiplying each value by a constant) is applied. Second, you will learn how to combine two or more random variables. This topic is critical as many of the statistical inference problems we will explore involve observing the difference between two random variables. You will learn how to describe the mean and standard deviation of the sum and difference of independent random variables as well as how to calculate probabilities of observing particular outcomes in these situations. There are a lot of formulas to keep straight in this section, so you may wish to keep your AP formula sheet handy!

### "Where Am I Going?"

**Learning Targets:**

- ____ I can describe the effects of transforming a random variable.
- ____ I can calculate and interpret the mean and standard deviation of the sum or difference of two random variables.
- ____ I can identify whether two random variables are independent.
- ____ I can find probabilities involving the sum or difference of independent Normal random variables.

### While You Read: Key Vocabulary and Concepts

linear transformation:

effect on a random variable of multiplying/dividing by a constant:

effect on a random variable of adding/subtracting a constant:

mean of the sum of random variables:

independent random variables:

variance of the sum of independent random variables:

mean of the difference of random variables:

variance of the difference of independent random variables:

**After You Read: "Where Am I Now?"**
**Check for Understanding**

***Concept 1: Linear Transformations***

We learned how linear transformations affect the shape, center, and spread of distributions of quantitative data back in Chapter 2. Similar rules apply to random variables. That is, when we multiply (or divide) each value of a random variable by a constant *b*, the shape of the probability distribution does not change. However, measures of center are multiplied (divided) by *b* and measures of spread are multiplied (divided) by |*b*|. When we add (or subtract) a constant *a* to each value of a random variable, the shape and spread of the probability distribution do not change. However, the measures of center will increase (or decrease) by *a*.

**Check for Understanding:** ____ *I can describe the effects of transforming a random variable.*

A carnival game involves tossing a ball into numbered baskets with the goal of having your ball land in a high-numbered basket. The probability distribution of X = value of the basket on a randomly selected toss.

| Value: | 0 | 1 | 2 | 3 |
|---|---|---|---|---|
| Probability: | 0.3 | 0.4 | 0.2 | 0.1 |

The expected value of X is 1.1 and its standard deviation is 0.0943.

Suppose it costs $2 to play and you earn $1.50 for each point earned on your toss. That is, if you land in a basket labeled "2," you will earn $3.00.

Define Y to be the amount of profit you make on a randomly selected toss. Describe the shape, center, and spread of the probability distribution of Y in the context of the situation.

***Concept 2: Combining Random Variables***

Many situations we'll encounter in later chapters involve two or more random variables. Understanding how to describe the center and spread of the probability distribution for the sum or difference of two random variables is an important skill to have. When given two independent random variables, we can describe the mean and standard deviation of the sum (or difference) of the random variables using the formulas in this section. Basically, when adding or subtracting two or more random variables (whether they are independent or not), the mean of the sum or difference of those random variables will be the sum or difference of their means. However, to describe the spread of the sum or difference of *independent* random variables, we must perform two steps. First, we find the *variance* of the sum or difference of two or more independent random variables by *adding* their variances. Then, we take the square root of the variance to find the standard deviation. A common mistake is to add standard deviations. Remember to *always add* variances! Never subtract and never combine standard deviations!

**Check for Understanding:** ____ *I can calculate and interpret the mean and standard deviation of the sum or difference of two random variables.*

Students in Mr. Costello's class are expected to check their homework in groups of 4 at the beginning of class each day. Students must check it as quickly as possible, one at a time. The means and standard deviations of the time it takes to check homework for the 4 students in one group are noted below. Assume their times are independent.

| | Mean | Standard Deviation |
|---|---|---|
| Alan | 1.4 min | 0.1 min |
| Barb | 1.2 min | 0.4 min |
| Corey | 0.9 min | 0.8 min |
| Doug | 1.0 min | 0.7 min |

a) If each student checks one after the other, what are the mean and standard deviation of the total time necessary for these four students to check their homework on a randomly chosen day?

b) Suppose Alan and Doug like to race to see who can check their homework faster. What are the mean and standard deviation for the difference between their times (Doug – Alan)? Interpret these values in the context of the situation.

### *Concept 3: Combining Normal Random Variables*

If our random variables of interest are Normally distributed, we can calculate the probability of observing particular outcomes using the skills we learned in Chapter 2. To do so, we rely on one important fact. When combining independent Normal random variables, the resulting distribution is also Normal! We can find the mean and standard deviation of the resulting distribution using the formulas we just learned. Then we can apply our knowledge from Chapter 2 to calculate and interpret probabilities about the situation.

**Check for Understanding:** ____ *I can find probabilities involving the sum or difference of independent Normal random variables.*

Mr. Molesky and Mr. Liberty are avid video game golfers. Both like to compare times to complete a particular course on their favorite game. Mr. Molesky's times are Normally distributed with a mean of 110 minutes and standard deviation of 10 minutes. Mr. Liberty's times are Normally distributed with mean 100 minutes and standard deviation 8 minutes.

a) Find the mean and standard deviation of the difference of their times (Molesky - Liberty). Assume their times are independent.

b) Find the probability that Mr. Molesky will finish his game before Mr. Liberty on any given day.

## Section 6.3: Binomial and Geometric Random Variables

### Before You Read: Section Summary

In the first two sections, you learned how to describe the probability distributions of discrete and continuous random variables as well as how to calculate probabilities for situations involving one or more random variables. In this section, you will focus on two special cases of discrete random variables: binomial and geometric. Binomial random variables count the number of successes that occur in a fixed number of independent trials of some chance process with a constant probability of success on each trial, while geometric random variables count the number of trials needed to get a success. Binomial random variables appear often on the AP exam, so you will want to pay particular attention to this topic. Again, try not to get bogged down in the formulas in this section. Familiarize yourself with the AP formula sheet and focus your efforts on being able to identify when to use binomial or geometric random variables and how to calculate and interpret probabilities involving them.

### "Where Am I Going?"

**Learning Targets:**

_____ I can determine whether the conditions for a binomial random variable have been met
_____ I can compute and interpret probabilities involving binomial distributions.
_____ I can calculate and interpret the mean and standard deviation of a binomial random variable.
_____ I can find probabilities involving geometric random variables.

### While You Read: Key Vocabulary and Concepts

binomial setting:

binomial random variable:

binomial coefficient:

binomial probability:

mean of a binomial random variable:

standard deviation of a binomial random variable:

Normal approximation for binomial distributions:

geometric setting:

geometric random variable:

geometric probability:

mean of a geometric random variable:

**After You Read: "Where Am I Now?"**
**Check for Understanding**

***Concept 1: Binomial Random Variables***

When we observe a fixed number of repeated trials of the same chance process, we are often interested in how many times a particular outcome occurs. This is the basis for a binomial setting. First, we are interested in outcomes that can be classified in one of two ways – success or failure. The particular outcome of interest is considered a success, while anything else is considered a failure. Next, each observed trial of the chance process must be independent of other trials. The number of trials we observe must be fixed in advance and the probability of success on each trial must be the same. If these conditions are met, we can use the binomial probability formula to determine the likelihood of observing a certain number of successes in a fixed number of trials of the binomial random variable:

$$P(X=k)=\binom{n}{k}p^k(1-p)^{n-k}$$

Note that the formula uses the multiplication rule for independent events from Chapter 5 in multiplying the probabilities of successes and failures across the fixed number of trials. However, the formula also considers the number of ways in which we can arrange those successes across our trials.

**Check for Understanding:** ____ *I can determine whether the conditions for a binomial random variable have been met and* ____ *I can compute and interpret probabilities involving binomial distributions.*

Recall that there are 4 suits—spades, hearts, clubs, and diamonds—in a standard deck of playing cards. Suppose you play a game in which you draw a card, record the suit, replace it, shuffle, and repeat until you have observed 10 cards. Define X = number of hearts observed.

a) Show that X is a binomial random variable.

b) Find the probability of observing fewer than 4 hearts in this game.

## *Concept 2: Mean and Standard Deviation of a Binomial Distribution and the Normal Approximation*

Like other discrete random variables, we can calculate the mean and standard deviation of binomial random variables. This will give us a better sense of what we'd expect to see in the long run as well as how much variability we can expect to observe in the observed number of successes. If a random variable X has a binomial probability distribution based on a chance process with $n$ trials each having probability of success $p$, we can calculate the mean of X by multiplying $np$. We can find the standard deviation of X by taking the square root of the product $np(1 - p)$. Modified versions of these formulas are useful when trying to make inferences about the proportion of successes in a population. If we take an SRS of size $n$ from a population (where $n$ is less than 10% of the size of the population), then we can use a binomial distribution to model the number of successes in the sample. Further, if $n$ is so large that both $np$ and $np(1 - p)$ are at least 10, we can use a Normal distribution to approximate binomial probabilities. As always, make sure you not only understand how to use the formulas, but also when to use them and how to interpret their results!

**Check for Understanding:** ____ *I can calculate and interpret the mean and standard deviation of a binomial random variable.*

Suppose 72% of students in the U.S. would give their teachers a positive rating if asked to score their effectiveness. A survey is conducted in which 500 students are randomly selected and asked to rate their teachers. Let X = the number of students in the sample who would give their teachers a positive rating.

a) Show that X is approximately a binomial random variable.

b) Use a Normal approximation to find the probability that 400 or more students would give their teacher a positive rating in this sample.

## *Concept 3: Geometric Random Variables*

In a binomial setting, we are interested in knowing how many successes will occur in a fixed number of trials. Sometimes we are interested in knowing how long it will take until a success occurs. When we perform independent trials of a chance process with the same probability of success on each trial, and record how long it takes to get a success, we have a geometric setting. We can describe the number of trials it takes to get a success using a geometric random variable. As with other random variables, we can describe the probability distribution of a geometric random variable and calculate its mean and standard deviation. Using what we learned in Chapter 5, we can calculate the probability of observing the first success on the $k^{th}$ trial by multiplying the probabilities of $(k-1)$ consecutive failures by the probability of a success:

$$P(Y=k)=(1-p)^{k-1}p.$$

**Check for Understanding:** ____ *I can find probabilities involving geometric random variables.*

Suppose 20% of Super Crunch cereal boxes contain a secret decoder ring. Let X = the number of boxes of Super Crunch that must be opened until a ring is found.

a) Show that X is a geometric random variable.

b) Find the probability that you will have to open 7 boxes to find a ring.

c) Find the probability that it will take fewer than 4 boxes to find a ring.

d) How many boxes would you expect to have to open to find a ring?

## Chapter Summary: Random Variables

In the last chapter we learned the basic definition and rules of probability. We continued our study of probability in this chapter by exploring situations that involve assigning a numerical value to each possible outcome of a chance process. The random variables that result form the foundation for inference procedures in later chapters. You learned how to calculate probabilities of events involving random variables as well as how to describe their probability distributions. You learned formulas to determine the mean and standard deviation of individual random variables as well as the combination of several independent random variables. Finally, you explored two special random variables – binomial and geometric – and learned how to calculate probabilities of events in binomial and geometric settings.

This chapter involved a lot of formulas, so you may want to familiarize yourself with the formula sheet provided on the AP exam. Like earlier chapters, you should focus less on memorizing formulas or calculator keystrokes and more on how to apply the formulas and interpret results. Make sure you understand how and when to apply each formula. More importantly, make sure you show your work when calculating probabilities so anyone reading your response understands exactly how you arrived at your answer!

### After You Read: "How Can I Close the Gap?"

Complete the vocabulary puzzle, multiple choice questions, and FRAPPY. Check your answers and your performance on each of the targets.

| Target | Got It! | Almost There | Needs Some Work |
|---|---|---|---|
| I can use a probability distribution to answer questions about possible values of a random variable. | | | |
| I can calculate and interpret the mean of a random variable. | | | |
| I can calculate and interpret the standard deviation of a random variable. | | | |
| I can describe the effects of transforming a random variable. | | | |
| I can calculate and interpret the mean and standard deviation of the sum or difference of two random variables. | | | |
| I can identify whether two random variables are independent. | | | |
| I can find probabilities involving the sum or difference of independent Normal random variables. | | | |
| I can describe the effects of transforming a random variable. | | | |
| I can determine whether the conditions for a binomial random variable have been met. | | | |
| I can compute and interpret probabilities involving binomial distributions. | | | |
| I can calculate and interpret the mean and standard deviation of a binomial random variable. | | | |
| I can find probabilities involving geometric random variables. | | | |

Did you check "Needs Some Work" for any of the targets? If so, what will you do to address your needs for those targets?

*Learning Plan:*

## Chapter 6 Multiple Choice Practice

**Directions.** *Identify the choice that best completes the statement or answers the question. Check your answers and note your performance when you are finished.*

1. A marketing survey compiled data on the number of cars in households. If $X$ = the number of cars in a randomly selected household, and we omit the rare cases of more than 5 cars, then $X$ has the following probability distribution:

| $X$ | 0 | 1 | 2 | 3 | 4 | 5 |
|---|---|---|---|---|---|---|
| $P(X)$ | 0.24 | 0.37 | 0.20 | 0.11 | 0.05 | 0.03 |

What is the probability that a randomly chosen household has at least two cars?

| | |
|---|---|
| A. | 0.19 |
| B. | 0.20 |
| C. | 0.29 |
| D. | 0.39 |
| E. | 0.61 |

2. What is the expected value of the number of cars in a randomly selected household?

| | |
|---|---|
| A. | 2.5 |
| B. | 0.1667 |
| C. | 1.45 |
| D. | 1 |
| E. | Can not be determined |

3. A dealer in Las Vegas selects 10 cards from a standard deck of 52 cards. Let $Y$ be the number of diamonds in the 10 cards selected. Which of the following best describes this setting?

| | |
|---|---|
| A. | $Y$ has a binomial distribution with $n$ = 10 observations and probability of success $p$ = 0.25. |
| B. | $Y$ has a binomial distribution with $n$ = 10 observations and probability of success $p$ = 0.25, provided the deck is shuffled well. |
| C. | $Y$ has a binomial distribution with $n$ = 10 observations and probability of success $p$ = 0.25, provided that after selecting a card it is replaced in the deck and the deck is shuffled well before the next card is selected. |
| D. | $Y$ has a geometric distribution with $n$ = 10 observations and probability of success $p$ = 0.25. |
| E. | $Y$ has a geometric distribution with n = 52 observations and probability of success $p$ = 0.25. |

4. In the town of Lakeville, the number of cell phones in a household is a random variable $W$ with the following probability distribution:

| Value $w_i$ | 0 | 1 | 2 | 3 | 4 | 5 |
|---|---|---|---|---|---|---|
| Probability $p_i$ | 0.1 | 0.1 | 0.25 | 0.3 | 0.2 | 0.05 |

The standard deviation of the number of cell phones in a randomly selected house is

| | |
|---|---|
| A. | 1.7475 |
| B. | 1.87 |
| C. | 2.5 |
| D. | 0.09 |
| E. | 2.9575 |

5. A random variable $Y$ has the following probability distribution:

| $Y$ | -1 | 0 | 1 | 2 |
|---|---|---|---|---|
| $P(Y)$ | $4C$ | $2C$ | 0.07 | 0.03 |

The value of the constant $C$ is:

| | |
|---|---|
| A. | 0.10. |
| B. | 0.15. |
| C. | 0.20. |
| D. | 0.25. |
| E. | 0.75. |

6. The variance of the sum of two random variables $X$ and $Y$ is

| | |
|---|---|
| A. | $\sigma_X + \sigma_Y$. |
| B. | $(\sigma_X)^2 + (\sigma_Y)^2$. |
| C. | $\sigma_X + \sigma_Y$, but only if $X$ and $Y$ are independent. |
| D. | $(\sigma_X)^2 + (\sigma_Y)^2$, but only if $X$ and $Y$ are independent. |
| E. | None of these. |

7. It is known that about 90% of the widgets made by Buckley Industries meet specifications. Every hour a sample of 18 widgets is selected at random for testing and the number of widgets that meet specifications is recorded. What is the approximate mean and standard deviation of the number of widgets meeting specifications?

| | |
|---|---|
| A. | $\mu = 1.62$; $\sigma = 1.414$ |
| B. | $\mu = 1.62$; $\sigma = 1.265$ |
| C. | $\mu = 16.2$; $\sigma = 1.62$ |
| D. | $\mu = 16.2$; $\sigma = 1.273$ |
| E. | $\mu = 16.2$; $\sigma = 4.025$ |

8. A raffle sells tickets for $10 and offers a prize of $500, $1000, or $2000. Let $C$ be a random variable that represents the prize in the raffle drawing. The probability distribution of $C$ is given below.

| Value $c_i$ | $0 | $500 | $1000 | $2000 |
|---|---|---|---|---|
| Probability $p_i$ | 0.60 | 0.05 | 0.13 | 0.22 |

The expected profit when playing the raffle is

| | |
|---|---|
| A. | $145. |
| B. | $585. |
| C. | $865. |
| D. | $635. |
| E. | $485. |

9. Let the random variable $X$ represent the amount of money Carl makes tutoring statistics students in the summer. Assume that $X$ is Normal with mean $240 and standard deviation $60. The probability is approximately 0.6 that, in a randomly selected summer, Carl will make less than about

| | |
|---|---|
| A. | $144 |
| B. | $216 |
| C. | $255 |
| D. | $30 |
| E. | $360 |

10. Which of the following random variables is geometric?

| | |
|---|---|
| A. | The number of phone calls received in a one-hour period |
| B. | The number of times I have to roll a six-sided die to get two 5s. |
| C. | The number of digits I will read beginning at a randomly selected starting point in a table of random digits until I find a 7. |
| D. | The number of 7s in a row of 40 random digits. |
| E. | All four of the above are geometric random variables. |

## Multiple Choice Answers

| Problem | Answer | Concept | Right | Wrong | Simple Mistake? | Need to Study More |
|---|---|---|---|---|---|---|
| 1 | D | Discrete Random Variable | | | | |
| 2 | C | Expected Value of Discrete Random Variables | | | | |
| 3 | C | Binomial Settings | | | | |
| 4 | A | Standard Deviation of Discrete Random Variables | | | | |
| 5 | B | Probability Distribution | | | | |
| 6 | D | Combining Random Variables | | | | |
| 7 | D | Binomial Approximations | | | | |
| 8 | B | Expected Value | | | | |
| 9 | C | Normal Approximations | | | | |
| 10 | C | Geometric Random Variables | | | | |

## FRAPPY! Free Response AP Problem, Yay!

The following problem is modeled after actual Advanced Placement Statistics free response questions. Your task is to generate a complete, concise response in 15 minutes. After you generate your response, view two example solutions and determine whether or not you feel they are "complete," "substantial," "developing," or "minimal." If they are not "complete," what would you suggest to the student who wrote them to increase their score? Finally, you will be provided with a rubric. Score your response and note what, if anything, you would do differently to increase your own score.

A recent study revealed that a new brand of mp3 player may need to be repaired up to 3 times during its ownership. Let *R* represent the number of repairs necessary over the lifetime of a randomly selected mp3 player of this brand. The probability distribution of the number of repairs necessary is given below.

| $r_i$ | 0 | 1 | 2 | 3 |
|---|---|---|---|---|
| $p_i$ | 0.4 | 0.3 | 0.2 | 0.1 |

(a) Compute and interpret the mean and standard deviation of *R*.

(b) Suppose we also randomly select a phone that may require repairs over its lifetime. The mean and standard deviation of the number of repairs for this brand of phone are 2 and 1.2, respectively. Assuming that the phone and mp3 player break down independently of each other, compute and interpret the mean and standard deviation of the total number of repairs necessary for the two devices.

(c) Each mp3 repair costs \$15 and each phone repair costs \$25. Compute the mean and standard deviation of the total amount you can expect to pay in repairs over the life of the devices.

**Student Response 1:**

a) mean = 0(0.4) + 1(0.3)+ 2(0.2) + 3(0.1) = 1
standard deviation = 1

We will have to repair our mp3 player exactly once or twice over its lifetime.

b) mean = 1 + 2 = 3
$1^2 + 1.2^2 = 2.44$   $\sqrt{(2.44)} = 1.562$ = standard deviation

We can expect to have to perform a total of about 3 repairs on the two devices. However, this can vary by up to 1.562 repairs or so over their lifetimes.

c) I would expect to pay \$15(1) + \$25(2) = \$65 in repairs over the lifetimes of the devices.

How would you score this response? Is it substantial? Complete? Developing? Minimal? Is there anything this student could do to earn a better score?

**Student Response 2:**

a) mean = 0(0.4) + 1(0.3)+ 2(0.2)+ 3(0.1) = 1
standard deviation = 1

This means we can expect to have to repair our mp3 player about once over its lifetime. But, we might have to repair it up to 2 times or maybe not at all. It is pretty unlikely we'd have to repair it three times.

b) mean = 1 + 2 = 3
standard deviation = 1 + 1.2 + 2.2

We can expect to have to perform a total of about 3 repairs on the two devices. However, this can vary by up to 2.2 repairs or so over their lifetimes.

c) The mean amount we can expect to pay in repairs is \$15(1) + \$25(2) = \$65. However, this amount will vary since we are not guaranteed to have to perform 3 repairs. The amount the cost can vary is $\sqrt{(15^2(1^2) + 25^2(1.2^2))}$ = \$33.54.

How would you score this response? Is it substantial? Complete? Developing? Minimal? Is there anything this student could do to earn a better score?

## Scoring Rubric

Use the following rubric to score your response. Each part receives a score of "Essentially Correct," "Partially Correct," or "Incorrect." When you have scored your response, reflect on your understanding of the concepts addressed in this problem. If necessary, note what you would do differently on future questions like this to increase your score.

## Intent of the Question

The goal of this question is to determine your ability to calculate and interpret the mean and standard deviation of a discrete random variable, combine random variables, and describe linear transformations of random variables.

## Solution

(a) Mean: $\mu_R = 0(0.4) + 1(0.3) + 2(0.2) + 3(0.1) = 1$

Standard deviation: $\sigma_R = \sqrt{(0-1)^2(0.4) + (1-1)^2(0.3) + (2-1)^2(0.2) + (3-1)^2(0.1)}$ = 1

We can expect to have to repair our mp3 player once over its lifetime, but that can vary on average by about 1 repair.

(b) If T = the total number of repairs across the two devices, the mean of T will be $\mu_T = 1 +$ 2 = 3 and the standard deviation will be $\sigma_T = \sqrt{1^2 + 1.2^2}$ = 1.562.

We can expect to have to perform 3 repairs, total, on our devices over their lifetime, but that amount can vary on average by about 1.562 repairs.

(c) The total amount we can expect to pay in repairs will be \$15(1) + \$25(2) = \$65.

The standard deviation will be $\sqrt{15^2(1^2) + 25^2(1.2)^2}$ = \$33.54.

We can expect to pay \$65 in repairs over the lifetime of the devices, but that can vary by an average of \$33.52.

**Scoring**:

Parts (a), (b), and (c) are scored as essentially correct (E), partially correct (P), or incorrect (I).

**Part (a)** is essentially correct if the mean and standard deviation are calculated correctly AND interpreted correctly.
Part (a) is partially correct if no interpretation or an incorrect interpretation is provided OR if only one calculation/interpretation is correct.

**Part (b)** is essentially correct if the mean and standard deviation are calculated correctly AND interpreted correctly.
Part (b) is partially correct if no interpretation or an incorrect interpretation is provided OR if only one calculation/interpretation is correct.

**Part (c)** is essentially correct if the mean and standard deviation are calculated correctly AND interpreted correctly.
Part (c) is partially correct if no interpretation or an incorrect interpretation is provided OR if only one calculation/interpretation is correct.

4 **Complete Response**
All three parts essentially correct

3 **Substantial Response**
Two parts essentially correct and one part partially correct

2 **Developing Response**
Two parts essentially correct and no parts partially correct
One part essentially correct and two parts partially correct
Three parts partially correct

1 **Minimal Response**
One part essentially correct and one part partially correct
One part essentially correct and no parts partially correct
No parts essentially correct and two parts partially correct

# Chapter 6: Random Variables

**Across**

2. The average of the squared deviations of the values of a variable from its mean.
5. Random variables are _____ if knowing whether an event in X has occurred tells us nothing about the occurrence of an event involving Y.
7. The probability _____ of a random variable gives its possible values and their probabilities.
10. The number of ways of arranging k successes among n observations is the binomial ____.
11. The sum or difference of independent Normal random variables follows a _____ distribution.
12. When you combine independent random variable, you always add these.
14. A linear _____ occurs when we add/subtract and multiply/divide by a constant.
16. An easy way to remember the requirements for a geometric setting.
17. This setting arises when we perform several independent trials of a chance process and record the number of times an outcome occurs.
18. The mean of a discrete random variable is also called the _____ value.

**Down**

1. A ____ variable takes numerical values that describe the outcomes of some chance proce
3. When n is large, we can use a Normal _____ determine probabilities for binomial settings.
4. A random variable that takes on all values in interval of numbers.
6. A random variable that takes a fixed set of possible values with gaps between.
8. A ____ setting arises when we perform independed trials of the same chance proces and record the number of trials until a particu outcome occurs.
9. An easy way to remember the requirements t a binmial setting.
13. Adding a constant to each value of a random variable has no effect on the shape or ____ c the distribution.
15. Multiplying each value of a random variable t constant has no effect on the ____ of the distribution.

# Chapter 7: Sampling Distributions

*"Statistics may be defined as 'a body of methods for making wise decisions in the face of uncertainty.'" W.A. Wallis*

## Chapter Overview

In chapters 1-3, you learned how to explore data. Chapter 4 introduced you to methods for producing data. Chapters 5 and 6 focused on the basics behind probability and random variables. In this chapter, you will learn the final piece necessary to study statistical inference – sampling distributions. The foundation of statistical inference lies in the concept of a sampling distribution. In order to make a conclusion about a population based on information from a sample, you need to be able to answer the question, "What results would I expect to see if I sampled repeatedly from my population of interest?" Sampling distributions provide an answer to that question and allow us to draw a conclusion about a population parameter based on an observed statistic from a sample. You will learn how to describe sampling distributions for sample proportions as well as sampling distributions for sample means. You will also learn an important theorem for sample means—the central limit theorem. The final chapters in the textbook will build upon your learning in this chapter to present formal methods for statistical inference. The better you understand sampling distributions, the easier your study of inference will be. Be sure to get a good grasp of the concepts in this chapter before moving on!

## Sections in this Chapter



## Plan Your Learning

Use the following *suggested* guide to help plan your reading and assignments. Note: your teacher may schedule a different pacing. Be sure to follow his or her instructions!

| **Read** | 7.1: pp 413-417 | 7.1: pp 417-428 | 7.2: pp 432-439 | 7.2: pp 442-448 |
|---|---|---|---|---|
| **Do** | 1, 3, 5, 7 | 9, 11, 13, 17-20 | 21-24, 27, 29, 33, 35, 37, 41 | 43-46, 49, 51, 53, 55 |

| **Read** | 7.3: pp 449-454 | Chapter Summary |
|---|---|---|
| **Do** | 57, 59, 61, 63, 65-68 | Multiple Choice FRAPPY! |

## Section 7.1: What is a Sampling Distribution?

### Before You Read: Section Summary

This section will introduce you to the big ideas behind sampling distributions. First, you will learn how to distinguish between population parameters and statistics derived from samples. Next, you will explore the fact that statistics vary from sample to sample. This simple fact is the reason we study sampling distributions. By describing the shape, center, and spread of the sampling distribution of a statistic, we can determine the critical information necessary to perform statistical inference in later chapters.

### "Where Am I Going?"

### Learning Targets:

- ____ I can distinguish between a parameter and a statistic.
- ____ I can define a sampling distribution.
- ____ I can distinguish between a population distribution, sampling distribution, and distribution of sample data.
- ____ I can determine whether a statistic is an unbiased estimator of a population parameter.
- ____ I can describe the relationship between sample size and the variability of an estimator.

### While You Read: Key Vocabulary and Concepts

parameter:

statistic:

sampling variability:

sampling distribution:

population distribution:

unbiased estimator:

variability of a statistic:

### After You Read: "Where Am I Now?" Check for Understanding

#### *Concept 1: Parameters and Statistics*

One of the most powerful skills we learn from statistics is the ability to answer a question about a population characteristic based on information gathered from a random sample. That is, we can use a statistic calculated from a sample to make a conclusion about a corresponding parameter in the population. However, we must note that the sample information we gather may differ somewhat from the population characteristic we are trying to measure. That's because the sample information would likely differ from sample to sample. This sample-to-sample variability poses a problem when we try to generalize our findings to the population. However, based on what we learned in the last chapter, we can view a sample statistic as a random variable. That is, while we have no way of predicting exactly what statistic value we will get from a sample, we know how those values will behave in repeated random sampling. With the probability distribution of this random variable in mind, we can use the sample statistic to estimate the population parameter.

**Check for Understanding:** ____ *I can distinguish between a parameter and a statistic.*

For each of the following situations, identify the population of interest, the parameter, and the statistic.

a) A medical researcher is interested in exploring the effects of a new medicine on blood pressure. 500 males with high blood pressure are randomly selected and given the new drug. After two weeks, their blood pressure is measured and the average arterial pressure is calculated.

b) A study is conducted to determine whether or not the dangerous activity of texting while driving is a common practice. 1500 16- to 24-year-olds are randomly selected and asked whether or not they text while driving. Of the 1500 drivers, 12% indicate they text while driving.

### *Concept 2: Describing Sampling Distributions*

To draw a conclusion about a population proportion *p*, we take a random sample and calculate the sample proportion $\hat{p}$. Likewise, to reach a conclusion about a population mean μ, we take a random sample and calculate the sample mean $\bar{x}$. Because of chance variation in random sampling, the values of our sample statistic will vary from sample to sample. The distribution of statistic values in all possible samples of the same size from a population is called the sampling distribution of the statistic. The sampling distribution describes the sampling variability and provides a foundation for performing inference. The spread of a sampling distribution is an important attribute as all inference calculations depend upon it! When trying to estimate a parameter, we want minimum sampling variability and no bias. Random sampling helps us avoid bias while larger samples help us minimize sampling variability.

**Check for Understanding:** ____ *I can distinguish between a population distribution, sampling distribution, and distribution of sample data.*

A breakfast cereal includes marshmallow shapes in the following distribution: 10% stars, 10% crescent moons, 20% rockets, 40% astronauts, 20% planets. We are interested in examining the proportion of rockets in a random sample of 2000 marshmallows from the cereal.

(a) Sketch the population distribution of marshmallow shapes.

(b) Suppose you were to collect a random sample of 2000 marshmallow shapes. Sketch the distribution of sample data you would expect to see. How many rockets would you expect to see in your sample?

(c) Now, suppose you collected many samples of the same size. Sketch the sampling distribution of the proportion of rockets you think you would see in the samples.

## Section 7.2: Sample Proportions

### Before You Read: Section Summary

The objective of some statistical applications is to reach a conclusion about a population proportion $p$. For example, we may try to estimate an approval rating through a survey, or test a claim about the proportion of defective light bulbs in a shipment based on a random sample. Since $p$ is unknown to us, we must base our conclusion on a sample proportion, $\hat{p}$. However, as we have noted, we know that the value of $\hat{p}$ will vary from sample to sample. The amount of variability will depend on the size of our sample. In this section, you will learn how to describe the shape, center, and spread of the sampling distribution of $\hat{p}$ in detail.

### "Where Am I Going?"

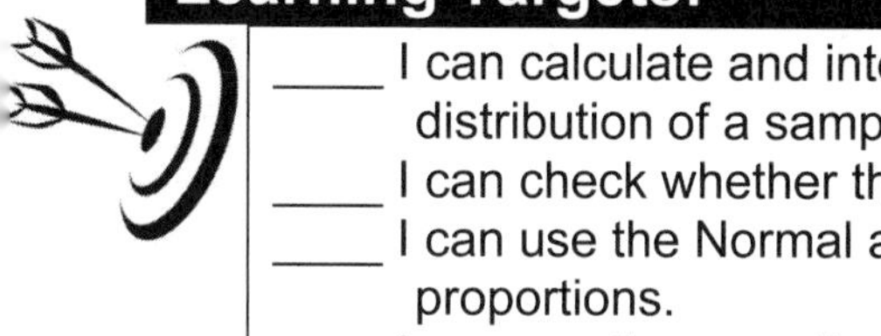

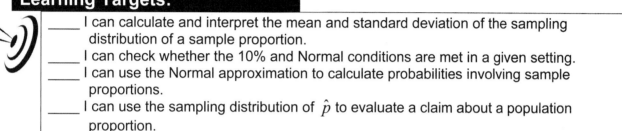

- ____ I can calculate and interpret the mean and standard deviation of the sampling distribution of a sample proportion.
- ____ I can check whether the 10% and Normal conditions are met in a given setting.
- ____ I can use the Normal approximation to calculate probabilities involving sample proportions.
- ____ I can use the sampling distribution of $\hat{p}$ to evaluate a claim about a population proportion.

### While You Read: Key Vocabulary and Concepts

sampling distribution of $\hat{p}$:

mean of the sampling distribution of $\hat{p}$:

standard deviation of the sampling distribution of $\hat{p}$:

Normal approximation for $\hat{p}$:

### After You Read: "Where Am I Now?" Check for Understanding

***Concept 1: The Sampling Distribution of $\hat{p}$***

If we take repeated samples of the same size $n$ from a population with a proportion of interest $p$, the sampling distribution of $\hat{p}$ will have the following characteristics:

1) The shape of the sampling distribution will become approximately Normal as the sample size $n$ increases. We can use Normal calculations if $np \geq 10$ and $n(1-p) \geq 10$.
2) The mean of the sampling distribution is $\mu_{\hat{p}} = p$.

3) The standard deviation of the sampling distribution is $\sigma_{\hat{p}} = \sqrt{\frac{p(1-p)}{n}}$

Note: The formula for the standard deviation is exactly correct only if we are sampling from an infinite population or *with replacement* from a finite population. When we are sampling without replacement from a finite population, the formula is approximately correct as long as the 10% condition is satisfied. That is, the sample size must be less than or equal to 10% of the population size.

**Check for Understanding:** ____ *I can calculate and interpret the mean and standard deviation of the sampling distribution of a sample proportion and* ____ *I can check whether the 10% and Normal conditions are met in a given setting.*

Suppose your job at a potato chip factory is to check each shipment of potatoes for quality assurance. Further, suppose that a truckload of potatoes contains 95% that are acceptable for processing. If more than 10% are found to be unacceptable in a random sample, you must reject the shipment. To check, you randomly select and test 250 potatoes. Let $\hat{p}$ be the sample proportion of unacceptable potatoes.

a) What is the mean of the sampling distribution of $\hat{p}$?

b) Check the 10% condition and calculate the standard deviation of the sampling distribution of $\hat{p}$.

c) Check the Normal condition and sketch the sampling distribution of $\hat{p}$. Based on this sketch, do you think it would be likely to reject the truckload based on a random sample of 250 potatoes? Why or why not?

### *Concept 2: Using the Normal Approximation for $\hat{p}$*

When the sample size *n* is large enough for *np* and *n(1-p)* to both be at least 10, the sampling distribution of $\hat{p}$ will be approximately Normal. In that case, we can use Normal calculations to determine the probability that an SRS will generate a value of $\hat{p}$ in a particular interval. This calculation is an important component of inference.

**Check for Understanding:** ____ *I can use the Normal approximation to calculate probabilities involving sample proportions and* ____ *I can use the sampling distribution of $\hat{p}$ to evaluate a claim about a population proportion.*

A phone company is interested in exploring marketing possibilities for a new smartphone for teenagers. They ask an SRS of 1000 high school students whether they own a smartphone. Suppose 65% of all high school students own a smartphone. What is the probability that the random sample selected by the company will result in a $\hat{p}$-value within 3 percentage points of the true population proportion? Show all your work!

## Section 7.3: Sample Means

### Before You Read: Section Summary

When the goal of a statistical application is to reach a conclusion about a population mean $\mu$ we must consider a sample mean $\bar{x}$. However, as we have noted, the value of $\bar{x}$ will vary from sample to sample. Like we observed with sample proportions, the amount of variability will depend on the size *n* of our sample. In this section, you will learn how to describe the shape, center, and spread of the sampling distribution of $\bar{x}$ in detail.

### "Where Am I Going?"

**Learning Targets:**

____ I can calculate and interpret the mean and standard deviation of the sampling distribution of a sample mean.

____ I can calculate probabilities involving a sample mean when the population distribution is Normal.

____ I can explain how the shape of the sampling distribution of $\bar{x}$ is related to the shape of the population distribution.

____ I can use the central limit theorem to help find probabilities involving a sample mean.

### While You Read: Key Vocabulary and Concepts

sampling distribution of $\bar{x}$:

mean of the sampling distribution of $\bar{x}$:

standard deviation of the sampling distribution of $\bar{x}$:

central limit theorem:

Normal condition for sample means:

### After You Read: "Where Am I Now?" Check for Understanding

#### *Concept 1: Sampling Distribution of $\bar{x}$*

If we take repeated random samples of the same size *n* from a population with mean $\mu$, the sampling distribution of $\bar{x}$ will have the following characteristics:

1) The shape of the sampling distribution will depend upon the shape of the population distribution. If the population is Normally distributed, the sampling distribution of $\bar{x}$ will be Normally distributed. If the population distribution is non-Normal, the sampling distribution of $\bar{x}$ will become more and more Normal as *n* increases.
2) The mean of the sampling distribution is $\mu_{\bar{x}} = \mu$.

3) The standard deviation of the sampling distribution is $\sigma_{\bar{x}} = \frac{\sigma}{\sqrt{n}}$

Note: The formula for the standard deviation is exactly correct only if we are sampling from an infinite population or *with replacement* from a finite population. When we are sampling without replacement from a finite population, the formula is approximately correct as long as the 10% condition is satisfied. That is, the sample size must be less than or equal to 10% of the population size.

**Check for Understanding:** ____ *I can calculate and interpret the mean and standard deviation of the sampling distribution of a sample mean and* ____ *I can calculate probabilities involving a sample mean when the population distribution is Normal.*

The times it takes 5$^{th}$ graders to complete a particular mathematics problem are Normally distributed with mean 2 minutes and standard deviation 0.8 minutes.

Find the probability that a randomly chosen 5$^{th}$ grader will take more than 2.5 minutes to complete the problem. Show your work.

Suppose you give the problem to an SRS of 20 students. Sketch the sampling distribution of $\bar{x}$. Then use this distribution to determine the probability that the mean time to complete the problem for the SRS of students is greater than 2.5 minutes. Show your work.

### *Concept 2: The Central Limit Theorem*

When the population is Normally distributed, we know that the sampling distribution of $\bar{x}$ will be Normally distributed, so we can use Normal calculations. However, most population distributions are not Normally distributed. If our sampling distribution is skewed or non-Normal in some other way, we cannot use Normal calculations to answer questions. Thankfully, a pretty remarkable fact about sample means helps us out: when the sample size $n$ is large, the shape of the sampling distribution of $\bar{x}$ will be approximately Normal no matter what the shape of the population distribution may be! For our purposes, we'll define "large" to be any sample that is at least 30. So, if $n \geq 30$, we can be safe in assuming that the sampling distribution of $\bar{x}$ will be approximately Normal and we can proceed to perform Normal calculations. If $n < 30$, we must consider the shape of the population distribution.

**Check for Understanding:** ____ *I can use the central limit theorem to help find probabilities involving a sample mean.*

The blood cholesterol level of adult men has mean 188 mg/dl and standard deviation 41 mg/dl. A SRS of 250 men is selected and the mean blood cholesterol level in the sample is calculated.

Sketch the sampling distribution of $\bar{x}$ and calculate the probability that the sample mean will be greater than 193.

## Chapter Summary: Sampling Distributions

This chapter introduced you to a key concept for inferential thinking – sampling distributions. Since we are interested in drawing conclusions about population proportions and means, it is important to know how sample proportions and means will behave in repeated random sampling. Being able to describe the sampling variability for our sample statistics will allow us to estimate and test claims about population parameters. This chapter provided us with some key facts about sample statistics that will help us as we begin our formal study of inference. First, statistics will vary from sample to sample. Second, if the sample size is large enough, we know that the distribution of sample statistic values will be approximately Normal. Third, the sampling distributions of $\hat{p}$ and $\bar{x}$ will be centered at $p$ and $\mu$, respectively . Finally, the variability of these sampling distributions can be computed (as long as the 10% condition is met) and used for inference calculations. This variability will decrease as the sample size increases, so bigger random samples are more desirable. One final, very important, fact about sample means was revealed in this chapter. When sampling from a Normal population, the sampling distribution of $\bar{x}$ will be Normal. However, as long as our sample size is at least 30, the shape of the sampling distribution of $\bar{x}$ will be approximately Normal no matter what the population distribution looks like! The central limit theorem is a powerful fact that will be revisited in the coming chapters.

Now that we have a grasp of the basic concept of sampling distributions, we are ready to begin the formal study of statistical inference. In the next chapter, we will use what we have learned in Chapter 7 to estimate population proportions and means with confidence.

### After You Read: "How Can I Close the Gap?"

Complete the vocabulary puzzle, multiple choice questions, and FRAPPY. Check your answers and your performance on each of the targets.

| Target | Got It! | Almost There | Needs Work |
|---|---|---|---|
| I can distinguish between a parameter and a statistic. | | | |
| I can define a sampling distribution. | | | |
| I can distinguish between a population distribution, sampling distribution, and distribution of sample data. | | | |
| I can determine whether a statistic is an unbiased estimator of a population parameter. | | | |
| I can describe the relationship between sample size and the variability of an estimator. | | | |
| I can calculate and interpret the mean and standard deviation of the sampling distribution of a sample proportion. | | | |
| I can check whether the 10% and Normal conditions are met in a given setting. | | | |
| I can use the Normal approximation to calculate probabilities involving sample proportions. | | | |
| I can use the sampling distribution of $\hat{p}$ to evaluate a claim about a population proportion. | | | |
| I can calculate and interpret the mean and standard deviation of the sampling distribution of a sample mean. | | | |
| I can calculate probabilities involving a sample mean when the population distribution is Normal. | | | |
| I can explain how the shape of the sampling distribution of $\bar{x}$ is related to the shape of the population distribution. | | | |
| I can use the central limit theorem to help find probabilities involving a sample mean. | | | |

Did you check "Needs Some Work" for any of the targets? If so, what will you do to address your needs for those targets?

*Learning Plan:*

## Chapter 7 Multiple Choice Practice

**Directions.** *Identify the choice that best completes the statement or answers the question. Check your answers and note your performance when you are finished.*

1. The variability of a statistic is described by

| | |
|---|---|
| A. | the spread of its sampling distribution. |
| B. | the amount of bias present. |
| C. | the vagueness in the wording of the question used to collect the sample data. |
| D. | probability calculations. |
| E. | the stability of the population it describes. |

2. Below are dot plots of the values taken by three different statistics in 30 samples from the same population. The true value of the population parameter is marked with an arrow.

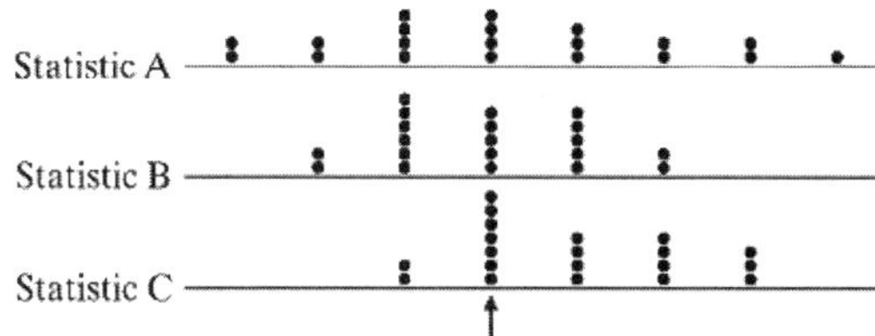


The statistic that has the largest *bias* among these three is

| | |
|---|---|
| A. | statistic A. |
| B. | statistic B. |
| C. | statistic C. |
| D. | A and B have similar bias, and it is larger than the bias of C. |
| E. | B and C have similar bias, and it is larger than the bias of A. |

3. According to a recent poll, 27% of Americans prefer to read their news in a physical newspaper instead of online. Let's assume this is the parameter value for the population. If you take a simple random sample of 25 Americans and let $\hat{p}$ = the proportion in the sample who prefer a newspaper, is the shape of the sampling distribution of $\hat{p}$ approximately Normal?

| | |
|---|---|
| A. | No, because $p < 0.50$ |
| B. | No, because $np = 6.75$ |
| C. | Yes, because we can reasonably assume there are more than 250 individuals in the population. |
| D. | Yes, because we took a simple random sample. |
| E. | Yes, because $n(1-p) = 18.25$ |

4. The time it takes students to complete a statistics quiz has a mean of 20.5 minutes and a standard deviation of 15.4 minutes. What is the probability that a random sample of 40 students will have a mean completion time greater than 25 minutes?

| | |
|---|---|
| A. | 0.9678 |
| B. | 0.0322 |
| C. | 0.0344 |
| D. | 0.3851 |
| E. | 0.6149 |

5. A fair coin (one for which both the probability of heads and the probability of tails are 0.5) is tossed 60 times. The probability that more than 1/3 of the tosses are heads is closest to

| | |
|---|---|
| A. | 0.9951. |
| B. | 0.33. |
| C. | 0.109. |
| D. | 0.09. |
| E. | 0.0049. |

6. The histogram below was obtained from data on 750 high school basketball games in a regional athletic conference. It represents the number of three-point baskets made in each game.

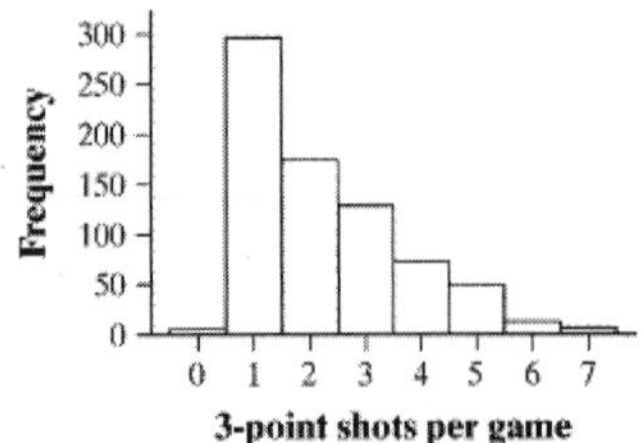


What is the range of sample sizes a researcher could take from this population without violating conditions required for performing Normal calculations with the sampling distribution of $\overline{x}$?

| | |
|---|---|
| A. | $0 \le n \le 30$ |
| B. | $30 \le n \le 50$ |
| C. | $30 \le n \le 75$ |
| D. | $30 \le n \le 750$ |
| E. | $75 \le n \le 750$ |

7. The incomes in a certain large population of college teachers have a normal distribution with mean \$60,000 and standard deviation \$5000. Four teachers are selected at random from this population to serve on a salary review committee. What is the probability that their average salary exceeds \$65,000?

| | |
|---|---|
| A. | 0.0228 |
| B. | 0.1587 |
| C. | 0.8413 |
| D. | 0.9772 |
| E. | essentially 0 |

8. A random sample of size 25 is to be taken from a population that is Normally distributed with mean 60 and standard deviation 10. The mean $\overline{x}$ of the observations in our sample is to be computed. The sampling distribution of $\overline{x}$

| | |
|---|---|
| A. | is Normal with mean 60 and standard deviation 10. |
| B. | is Normal with mean 60 and standard deviation 2. |
| C. | is approximately Normal with mean 60 and standard deviation 2. |
| D. | has an unknown shape with mean 60 and standard deviation 10. |
| E. | has an unknown shape with mean 60 and standard deviation 2. |

9. The scores of individual students on a college entrance examination have a left-skewed distribution with mean 18.6 and standard deviation 6.0. At Millard North High School, 36 seniors take the test. The sampling distribution of mean scores for random samples of 36 students is

| | |
|---|---|
| A. | approximately Normal. |
| B. | symmetric and mound-shaped, but non-Normal. |
| C. | skewed right. |
| D. | neither Normal nor non-normal. It depends on the particular 36 students selected. |
| E. | exactly Normal. |

10. The distribution of prices for home sales in Minnesota is skewed to the right with a mean of \$290,000 and a standard deviation of \$145,000. Suppose you take a simple random sample of 100 home sales from this (very large) population. What is the probability that the mean of the sample is above \$325,000?

| | |
|---|---|
| A. | 0.0015 |
| B. | 0.0027 |
| C. | 0.0079 |
| D. | 0.4046 |
| E. | 0.4921 |

## Multiple Choice Answers

| Problem | Answer | Concept | Right | Wrong | Simple Mistake? | Need to Study More |
|---|---|---|---|---|---|---|
| 1 | A | Sampling Variability | | | | |
| 2 | C | Bias and Variability | | | | |
| 3 | B | Normality Condition | | | | |
| 4 | B | Normal Probability Calculation | | | | |
| 5 | A | Normal Probability Calculation | | | | |
| 6 | C | 10% Condition and CLT | | | | |
| 7 | A | Normal Probability Calculation | | | | |
| 8 | B | Sampling Distribution for Means | | | | |
| 9 | A | Sampling Distribution for Means | | | | |
| 10 | C | Normal Probability Calculation | | | | |

## FRAPPY! Free Response AP Problem, Yay!

The following problem is modeled after actual Advanced Placement Statistics free response questions. Your task is to generate a complete, concise response in 15 minutes. After you generate your response, view two example solutions and determine whether or not you feel they are "complete," "substantial," "developing" or "minimal." If they are not "complete," what would you suggest to the student who wrote them to increase their score? Finally, you will be provided with a rubric. Score your response and note what, if anything, you would do differently to increase your own score.

A television producer must schedule a selection of paid advertisements during each hour of programming. The lengths of the advertisements are Normally distributed with a mean of 28 seconds and standard deviation of 5 seconds. During each hour of programming, 45 minutes are devoted to the program and 15 minutes are set aside for advertisements. To fill in the 15 minutes, the producer randomly selects 30 advertisements.

a) Describe the sampling distribution of the sample mean length for random samples of 30 advertisements.

b) If 30 advertisements are randomly selected, what is the probability that the total time needed to air them will exceed the 15 minutes available? Show your work.

**Student Response 1:**

**(a)** The sampling distribution will have a mean of 28 seconds and a standard deviation of 5/√(28) = 0.945 seconds.

**(b)** z=(30-28)/0.945 = 2.12. The probability the total time will exceed the 15 available minutes is 1 - 0.9826 = 0.0174.

How would you score this response? Is it substantial? Complete? Developing? Minimal? Is there anything this student could do to earn a better score?

**Student Response 2:**

**(a)** Since the sample size is only 28, we can't use the central limit theorem. However, because the advertisement times are Normally distributed, the sampling distribution of the average time will be Normal. The mean will be 28 seconds and the standard deviation is 5/√(30) = 0.913 seconds.

**(b)** Since 15 minutes = 900 seconds, the average time for the 30 ads can not exceed 900/30 = 30 seconds. Using the information from part (a), z = (30-28) / 0.913 = 2.19. Therefore, the probability the ads will exceed 900 seconds is 1 – 0.9857 = 0.0143. This is not very likely, so the producer should not be concerned about running over if he or she selects 30 ads randomly.

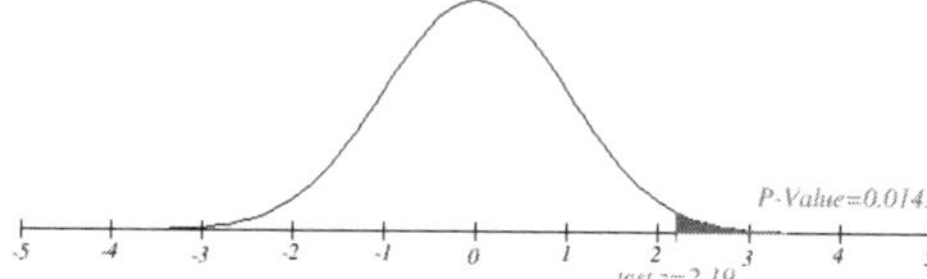


How would you score this response? Is it substantial? Complete? Developing? Minimal? Is there anything this student could do to earn a better score?

**Scoring Rubric**

Use the following rubric to score your response. Each part receives a score of "Essentially Correct," "Partially Correct," or "Incorrect." When you have scored your response, reflect on your understanding of the concepts addressed in this problem. If necessary, note what you would do differently on future questions like this to increase your score.

**Intent of the Question**

The goal of this question is to determine your ability to describe the sampling distribution of a sample mean and use it to perform a probability calculation.

**Solution**

(a) The sampling distribution of the sample mean length for random samples of 30 advertisements has mean 28 seconds and standard deviation $\sigma_{\bar{x}} = \dfrac{5}{\sqrt{30}} = 0.913$ seconds. Because we are told that the population of advertisement lengths is Normally distributed, the shape of the sampling distribution will be Normal.

(b) The probability that a random sample of 30 advertisements will exceed the allotted time is equivalent to the probability that the sample mean length of the 30 advertisements is greater than 900/30 = 30 seconds. In part (a), we determined that the sampling distribution is N(28, 0.913). Therefore,

$$P(\bar{x} > 30) = P\left(Z > \frac{30-28}{0.913}\right) = P(Z > 2.19) = 1 - 0.9857 = 0.0143$$

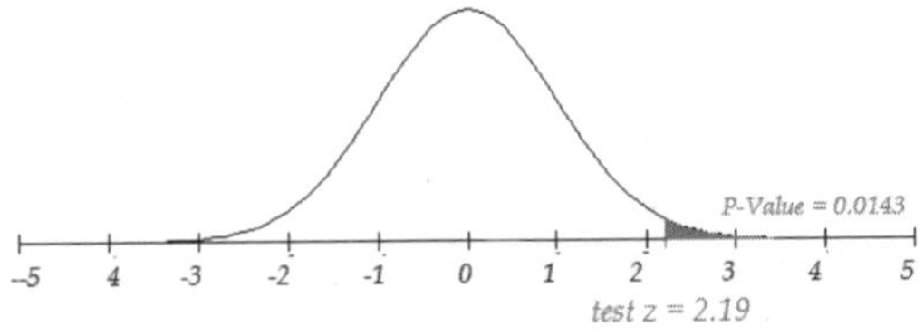

There is a 1.43% chance the randomly selected advertisements will exceed the allotted time.

**Scoring**:

Parts (a), (b), and (c) are scored as essentially correct (E), partially correct (P), or incorrect (I).

**Part (a)** is essentially correct if the response correctly identifies the shape (Normal), center (mean=28 seconds) and spread (standard deviation = 0.913 seconds) of the sampling distribution. The calculation of the standard deviation should be shown to earn an essentially correct score. Part (a) is partially correct if the solution only identifies 2 of the 3 components correctly or correctly identifies the standard deviation, but fails to show the calculation.

**Part (b)** is essentially correct if the response sets up and performs a correct probability calculation. Part (b) is partially correct if the response includes a correctly set up calculation, but fails to calculate the correct value OR if it sets up an incorrect, but plausible, calculation but carries it through correctly.

4 **Complete Response**
Both parts essentially correct

3 **Substantial Response**
One part essentially correct and one part partially correct

2 **Developing Response**
Both parts partially correct
One part essentially correct

1 **Minimal Response**
One part partially correct

# Chapter 7: Sampling Distributions

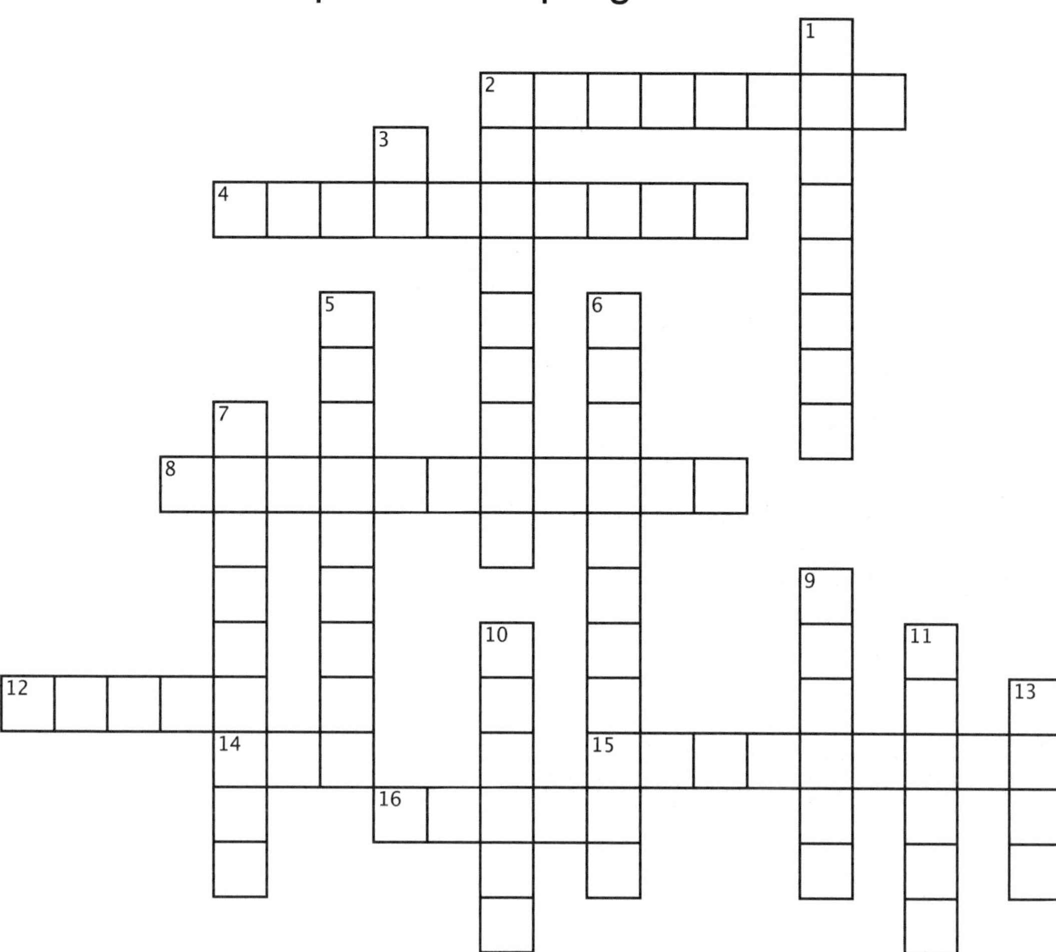


## ross

_____ distribution: the distribution of values taken by the statistic in all possible samples of the same size from the population

______ distribution: the distribution of all values of a variable in the population

_____ of a statistic is described by the spread of the sampling distribution

Greek letter used for the population standard deviation

the Normal approximation for the sampling distribution of a sample proportion can be used when both the number of successes and failures are greater than _____

sampling distributions and sampling variability provide the foundation for performing ______

central _____ theorem tells us if the sample size is large, the sampling distribution of the sample mean is approximately Normal, regardless of the shape of the population

## Down

1. a statistic is an _____ estimator if the mean of the sampling distribution is equal to the true value of the parameter being estimated.
2. a number, computed from sample data, that estimates a parameter
3. Greek letter used for the population mean
5. standard ______ : measure of spread of a sampling distribution
6. sampling ______ notes the value of a statistic may be different from sample to sample
7. a number that describes a population
9. the rule of thumb for using the central limit theorem - the sample size should be greater than ______
10. when the sample size is large, the sampling distribution of a sample proportion is approximately ______
11. to draw a conclusion about a population parameter, we can look at information from a _____ sample
13. center of a sampling distribution

# Chapter 8: Estimating with Confidence

*"Do not put your faith in what statistics say until you have carefully considered what they do not say." William W. Watt*

## Chapter Overview

Now that you have learned the basics of probability and sampling distributions, you are ready to begin your formal study of inference. In this chapter, you will use what you have learned about sampling distributions to construct confidence intervals for population proportions and population means. In later chapters, you'll learn how to determine confidence intervals for paired data, differences in means and proportions, and slopes of regression lines. Each of those procedures uses the same approach as the intervals you'll construct in this chapter, so you'll want to get a solid foundation in the basics here!

## Sections in this Chapter

**Section 8.1**: Confidence Intervals: The Basics
**Section 8.2**: Estimating a Population Proportion
**Section 8.3**: Estimating a Population Mean

## Plan Your Learning

Use the following *suggested* guide to help plan your reading and assignments. Note: your teacher may schedule a different pacing. Be sure to follow his or her instructions!

| **Read** | 8.1: pp 467-478 | 8.1: pp 478-481 | 8.2: pp 484-495 | 8.3: pp 499-511 |
|---|---|---|---|---|
| **Do** | 5, 7, 9, 11, 13 | 17, 19-24, 27, 31, 33 | 35, 37, 41, 43, 47 | 49-52, 55, 57, 59, 63 |

| **Read** | 8.3: pp 511-517 | Chapter Summary |
|---|---|---|
| **Do** | 65, 57, 71, 73, 75-78 | Multiple Choice FRAPPY! |

## Section 8.1: Confidence Intervals: The Basics

### Before You Read: Section Summary

In this section you will be introduced to the basic ideas behind constructing and interpreting a confidence interval for a population parameter. You will learn how we can take a point estimate for a population parameter and use what we know about sampling variability to construct an interval of plausible values for the parameter. You will focus on the big ideas in this section. You will want to make sure you understand the different components of a confidence interval as well as the correct interpretation of both the interval and the confidence level. The next two sections will build upon the concepts presented here and focus on the details for estimating proportions and means.

### "Where Am I Going?"

### Learning Targets:

- ____ I can interpret a confidence level.
- ____ I can interpret a confidence interval in context.
- ____ I can explain that a confidence interval gives a range of plausible values for the parameter.
- ____ I can explain why each of the three inference conditions – random, Normal, and independent – is important.
- ____ I can explain how issues like nonresponse, undercoverage, and response bias can influence the interpretation of a confidence interval.
- ____ I can explain how sample size and level of confidence *C* affect the margin of error of a confidence interval.

### While You Read: Key Vocabulary and Concepts

point estimator:

point estimate:

confidence interval:

margin of error:

confidence level *C*:

critical value:

conditions for constructing a confidence interval:

**After You Read: "Where Am I Now?"**
**Check for Understanding**

### *Concept 1: The Idea of a Confidence Interval*

When our goal is to estimate a population parameter, we often must rely on a sample statistic to provide a "point estimate." However, as we learned in the last chapter, that estimate will vary from sample to sample. A confidence interval takes that variation in to account to provide an interval of plausible values, based on the statistic, for the true parameter. All confidence intervals have two main components: an interval based on the estimate that includes a margin of error and a confidence level *C* that reports the success rate of the method used to construct the interval in capturing the parameter in repeated constructions. For example, "C% confident" means C% of all samples of the same size from the population of interest would yield an interval that captures the true parameter. We can then interpret the interval itself to say "We are C% confident that the interval from a to b captures the true value of the population parameter."

**Check for Understanding:** ____ *I can interpret a confidence level,* ____ *I can interpret a confidence interval in context, and* ____ *I can explain that a confidence interval gives a range of plausible values for the parameter.*

How much do the volumes of bottles of water vary? A random sample of 50 "20 oz." water bottles is collected and the contents are measured. A 90% confidence interval for the population mean $\mu$ is 19.10 to 20.74.

a) Interpret the confidence interval in context.

b) Interpret the confidence level in context.

c) Based on this interval, what can you say about the contents of the bottles in the sample? What can you say about the contents of bottles in the population?

### *Concept 2: Constructing a Confidence Interval*

To construct a confidence interval, we must work through three steps. First, you MUST check that the conditions for constructing the interval are met. That is, we must be assured that the data come from a random sample or randomized experiment. The sampling distribution of the statistic must be approximately Normal. And, the individual observations must be independent (which means checking the 10% condition if we're sampling without replacement from a finite population). Second, we construct the interval using the formula

statistic ± (critical value) (standard deviation of the statistic) where the critical value is determined based on the confidence level *C* and the standard deviation is based on the sampling distribution of the statistic. Finally, we interpret the interval using the language we learned earlier in this section.

Our goal with confidence intervals is to provide as precise an estimate as possible. That is, we wish to construct a narrow interval that we are confident captures the parameter of interest. We can achieve this in two ways: by decreasing our confidence or by increasing our sample size.

**Check for Understanding:** ____ *I can explain why each of the three inference conditions – random, Normal, and independent – is important and* ____ *I can explain how issues like nonresponse, undercoverage, and response bias can influence the interpretation of a confidence interval.*

A large company is interested in developing a new bake ware product for consumers. In an effort to determine baking habits of adults, a researcher selects a random sample of 50 addresses in a large, Midwestern, metropolitan area. She calls each selected home in the late-morning to collect information on their baking habits. The proportion of adults who bake at least twice a week is calculated and a 90% confidence interval is constructed.

Discuss whether or not each of the conditions for constructing a confidence interval has been met. If any have not been met, discuss the implications on the interpretation of the interval.

## Section 8.2: Estimating a Population Proportion

### Before You Read: Section Summary

In the last section, you learned the basic ideas behind confidence intervals. In the next two sections, you will learn how to construct and interpret confidence intervals for proportions and means. You will start by constructing them for proportions, focusing on the application of the four-step process to the procedure.

### "Where Am I Going?"

**Learning Targets:**

____ I can construct and interpret a confidence interval for a population proportion.
____ I can determine critical values for calculating a confidence interval.
____ I can determine the sample size necessary to obtain a level *C* confidence interval for a population proportion with a specified margin of error.

### While You Read: Key Vocabulary and Concepts

conditions for estimating *p*:

standard error:

confidence interval for *p*:

sample size for a desired margin of error:

### After You Read: "Where Am I Now?"
### Check for Understanding

#### *Concept 1: Conditions for Estimating p*

When constructing a confidence interval for *p*, it is critical that you begin by checking that the conditions are met. First, check to make sure that the sample was randomly selected or there was random assignment in an experiment. Because the construction of the interval is based on the sampling distribution of $\hat{p}$, next you must ensure that the condition for Normality is met. That is, check to see that $n\hat{p}$ and $n(1-\hat{p})$ are both at least 10. Finally, check for independence of measurements. If there is sampling without replacement, verify that the population of interest is at least 10 times as large as the sample. If all three of these conditions are met, you can safely proceed to construct and interpret a confidence interval for a population proportion *p*.

## *Concept 2: Constructing a Confidence Interval for p*

To construct a confidence interval for a population proportion *p*, you should follow the four step process introduced in Chapter 1.

- **State** the parameter you want to estimate and at what confidence level.
- **Plan** which confidence interval you will construct and verify that the conditions have been met.
- **Do** the actual construction of the interval using the basic idea from Section 8.1

$$\hat{p} \pm z^* \sqrt{\frac{\hat{p}(1-\hat{p})}{n}}$$

  where $z^*$ is the critical value for the standard Normal curve with area *C* between $-z^*$ and $z^*$.
- **Conclude** by interpreting the interval in the context of the problem.

**Check for Understanding:** ____ *I can construct and interpret a confidence interval for a population proportion and ____ I can determine critical values for calculating a confidence interval.*

According to a recent study, not everyone can roll their tongue. A researcher observed a random sample of 300 adults and found 68 who could roll their tongue. Use the four-step process to construct and interpret a 90% confidence interval for the true proportion of adults who can roll their tongue.

### *Concept 3: Choosing the Sample Size*

As noted in section 8.1, our goal is to estimate the parameter as precisely as possible. We want high confidence and a low margin of error. To achieve that, we can determine how large a sample size is necessary before proceeding with the data collection. To calculate the sample size necessary to achieve a set margin of error at a confidence level C, we simply solve the following inequality for *n*:

$$z^* \sqrt{\frac{\hat{p}(1-\hat{p})}{n}} \leq ME$$

where $\hat{p}$ is estimated based on a previous study or set to 0.5 to maximize the possible margin of error.

**Check for Understanding:** ____ *I can determine the sample size necessary to obtain a level C confidence interval for a population proportion with a specified margin of error.*

A researcher would like to estimate the proportion of adults who can roll their tongues. However, unlike the previous example, she'd like the estimate to be within 2% at a 95% confidence level. How large a sample is needed?

## Section 8.3: Estimating a Population Mean

### Before You Read: Section Summary

In this section, you will continue your study of confidence intervals by learning how to construct and interpret a confidence interval for a mean. While the overall procedure is identical to that for a proportion, there is one major difference. When dealing with means and unknown population standard deviations, we must use a new distribution to determine critical values. You will be introduced to the *t*-distributions and learn how to use them to construct a confidence interval for a population mean.

### "Where Am I Going?"

**Learning Targets:**

____ I can construct and interpret a confidence interval for a population mean.
____ I can determine the sample size required to obtain a level C confidence interval for a population mean with a specified margin of error.
____ I can determine sample statistics from a confidence interval.

### While You Read: Key Vocabulary and Concepts

one-sample *z*-interval for a population mean:

*t*-distribution:

degrees of freedom:

standard error of the sample mean:

one-sample *t*-interval for a population mean:

conditions for inference about a population mean:

robust procedures:

### After You Read: "Where Am I Now?" Check for Understanding

#### *Concept 1: Conditions for Estimating* μ

Like proportions, when constructing a confidence interval for $\mu$, it is critical that you begin by checking that the conditions are met. First, check to make sure the sample was randomly selected or there was random assignment in an experiment. Because the construction of the interval is based on the sampling distribution of $\bar{x}$, next ensure that the condition for Normality is met. That is, check to see that the population distribution is Normal OR the sample size is at least 30. Finally, check for independence of measurements. If sampling without replacement was used, verify that the population of interest is at least 10 times as large as the sample. If all three of these conditions are met, you can safely proceed to construct and interpret a confidence interval for a population mean $\mu$.

### *Concept 2: t-Distributions*

When the population standard deviation is unknown, we can no longer model the test statistic with the Normal distribution. Therefore, we can't use critical $z^*$ values to determine the margin of error in our confidence interval. Fortunately, it turns out that when the Normal condition is met, the test statistic calculated using the sample standard deviation $s_x$ has a distribution similar in appearance to the Normal distribution, but with more area in the tails. That is, the statistic

$$t = \frac{\bar{x} - \mu}{s_x / \sqrt{n}}$$

has the *t*-distribution with ($n$-1) degrees of freedom. As the sample size and, subsequently, the degrees of freedom increase, the *t* distribution approaches the standard Normal distribution more and more closely. We calculate standardized *t* values the same way we calculate *z* values. However, we must refer to a *t* table and consider degrees of freedom when determining critical values.

The *t*-procedures are fairly robust against slight departures from Normality in the population distribution. However, you should exercise caution in using *t*-procedures when there is evidence of strong skewness or outliers in the sample data.

**Check for Understanding:**

Use the *t* table to determine the critical value $t^*$ that you would use for a confidence interval for a population mean $\mu$ in the following situations.

a) An 80% confidence interval from a sample with size $n$ = 19

b) A 95% confidence interval from 248 degrees of freedom

c) A 99% confidence interval for a sample with size $n$ = 30

### *Concept 3: Constructing a Confidence Interval for μ*

To construct a confidence interval for a population mean $\mu$ when the population standard deviation is unknown, you should follow the four step process introduced in Chapter 1.

- **State** the parameter you want to estimate and at what confidence level.
- **Plan** which confidence interval you will construct and verify that the conditions have been met.
- **Do** the actual construction of the interval using the basic idea from Section 8.1

  $$\bar{x} \pm t^* \frac{s_x}{\sqrt{n}}$$

  where $t^*$ is the critical value for the $t_{n-1}$ distribution with area *C* between $-t^*$ and $t^*$.
- **Conclude** by interpreting the interval in the context of the problem.

**Check for Understanding:** ____ *I can construct and interpret a confidence interval for a population mean.*

The amount of sugar in soft drinks is increasingly becoming a concern. To test sugar content, a researcher randomly sampled 8 soft drinks from a particular manufacturer and measured the sugar content in grams/serving. The following data were produced:

26 31 23 22 11 22 14 31

Use these data to construct and interpret a 95% confidence interval for the mean amount of sugar in this manufacturer's soft drinks.

## *Concept 4: Choosing the Sample Size*

Similar to what we did with proportions, to calculate the sample size necessary to achieve a set margin of error for a population mean $\mu$ at a confidence level *C*, we simply solve the following inequality for *n*:

$$z^* \frac{\sigma}{\sqrt{n}} \leq ME$$

where $\sigma$ is estimated based on a previous study.

**Check for Understanding:** ____ *I can determine the sample size required to obtain a level C confidence interval for a population mean with a specified margin of error.*

A researcher would like to estimate the mean amount of time it takes to accomplish a particular task. A previous study indicates the time required varies in the population with a standard deviation of 4 seconds. He would like to estimate the true mean time within 0.5 seconds at 95% confidence. How large a sample is needed?

## Chapter Summary: Estimating with Confidence

Statistical inference is the practice of drawing a conclusion about a population based on information gathered from a sample. This chapter introduced us to the practice of estimating a parameter based on a statistic. The underlying logic for confidence intervals is the same whether we are estimating a proportion or a mean. By using what we learned about sampling distributions, we can construct an interval around a point estimate that we are confident captures the parameter of interest. The confidence level itself tells what would happen if we used the construction method for the interval many times for samples of the same size. It is basically the capture rate for all of the constructed intervals. So, when we build a level C interval, we can interpret it by saying "We are C% confident the interval from a to b captures the true parameter of interest."

In the next chapter, we will learn how to test a claim about a parameter. Make sure you continue to practice confidence intervals, though, as they are just as important as the significance tests you are about to learn!

### After You Read: "How Can I Close the Gap?"

Complete the vocabulary puzzle, multiple choice questions, and FRAPPY. Check your answers and on your performance on each of the targets.

| Target | Got It! | Almost There | Needs Some Work |
|---|---|---|---|
| I can interpret a confidence level in context. | | | |
| I can interpret a confidence interval in context. | | | |
| I can explain that a confidence interval gives a range of plausible values for the parameter. | | | |
| I can explain why each of the three inference conditions – random, Normal, and independent – is important. | | | |
| I can explain how issues like nonresponse, undercoverage, and response bias can influence the interpretation of a confidence interval. | | | |
| I can construct and interpret a confidence interval for a population proportion. | | | |
| I can determine critical values for calculating a confidence interval. | | | |
| I can determine the sample size necessary to obtain a level C confidence interval for a population proportion with a specified margin of error. | | | |
| I can explain how sample size and level of confidence C affect the margin of error of a confidence interval. | | | |
| I can construct and interpret a confidence interval for a population mean. | | | |
| I can determine the sample size required to obtain a level C confidence interval for a population mean with a specified margin of error. | | | |
| I can determine sample statistics from a confidence interval. | | | |

Did you check "Needs Some Work" for any of the targets? If so, what will you do to address your needs for those targets?

*Learning Plan:*

## Chapter 8 Multiple Choice Practice

**Directions.** *Identify the choice that best completes the statement or answers the question. Check your answers and note your performance when you are finished.*

1. Gallup Poll interviews 1600 people. Of these, 18% say that they jog regularly. A news report adds: "The poll had a margin of error of plus or minus three percentage points." You can safely conclude that

| | |
|---|---|
| A. | 95% of all Gallup Poll samples like this one give answers within ±3% of the true population value. |
| B. | the percent of the population who jog is certain to be between 15% and 21%. |
| C. | 95% of the population jog between 15% and 21% of the time. |
| D. | we can be 3% confident that the sample result is true. |
| E. | if Gallup took many samples, 95% of them would find that 18% of the people in the sample jog. |

2. An agricultural researcher plants 25 plots with a new variety of corn. A 90% confidence interval for the average yield for these plots is found to be 162.72 ± 4.47 bushels per acre. Which of the following is the correct interpretation of the interval?

| | |
|---|---|
| A. | There is a 90% chance the interval from 158.28 to 167.19 captures the true average yield. |
| B. | 90% of sample average yields will be between 158.28 and 167.19 bushels per acre. |
| C. | We are 90% confident the interval from 158.28 to 167.19 captures the true average yield. |
| D. | 90% of the time, the true average yield will fall between 158.28 and 167.19. |
| E. | We are 90% confident the true average yield is 162.72. |

3. I collect a random sample of size *n* from a population and from the data collected compute a 95% confidence interval for the mean of the population. Which of the following would produce a wider confidence interval, based on these same data?

| | |
|---|---|
| A. | Use a larger confidence level. |
| B. | Use a smaller confidence level. |
| C. | Use the same confidence level, but compute the interval *n* times. Approximately 5% of these intervals will be larger. |
| D. | Increase the sample size. |
| E. | Nothing can ensure that you will get a larger interval. One can only say the chance of obtaining a larger interval is 0.05. |

4. A marketing company discovered the following problems with a recent poll:
I. Some people refused to answer questions
II. People without telephones could not be in the sample
III. Some people never answered the phone in several calls
Which of these sources is included in the ±2% margin of error announced for the poll?

| | |
|---|---|
| A. | Only source I. |
| B. | Only source II. |
| C. | Only source III. |
| D. | All three sources of error. |
| E. | None of these sources of error. |

5. You are told that the proportion of those who answered "yes" to a poll about internet use is 0.70, and that the standard error is 0.0459. The sample size

| | |
|---|---|
| A. | is 50. |
| B. | is 99. |
| C. | is 100. |
| D. | is 200. |
| E. | cannot be determined from the information given. |

6. The standardized test scores of 16 students have mean $\bar{x}$ = 200 and standard deviation $s$ = 20. What is the standard error of $\bar{x}$?

| | |
|---|---|
| A. | 20 |
| B. | 10 |
| C. | 5 |
| D. | 1.25 |
| E. | 0.80 |

7. A newspaper conducted a statewide survey concerning the 2008 race for state senator. The newspaper took a random sample (assume it is an SRS) of 1200 registered voters and found that 620 would vote for the Republican candidate. Let $p$ represent the proportion of registered voters in the state that would vote for the Republican candidate. A 90% confidence interval for $p$ is

| | |
|---|---|
| A. | 0.517 ± 0.014. |
| B. | 0.517 ± 0.022. |
| C. | 0.517 ± 0.024. |
| D. | 0.517 ± 0.028. |
| E. | 0.517 ± 0.249. |

8. After a college's football team once again lost a football game to the college's arch rival, the alumni association decided to conduct a survey to see if alumni were in favor of firing the coach. Let $p$ represent the proportion of all living alumni who favor firing the coach. Which of the following is the smallest sample size needed to guarantee an estimate that's within 0.05 of $p$ at a 95% confidence level?

| | |
|---|---|
| A. | 269 |
| B. | 385 |
| C. | 538 |
| D. | 768 |
| E. | 1436 |

9. An SRS of 100 postal employees found that the average time these employees had worked for the postal service was $\bar{x}$ = 7 years with standard deviation $s_x$ = 2 years. Assume the distribution of the time the population of employees has worked for the postal service is approximately Normal. A 95% confidence interval for the mean time $\mu$ the population of postal service employees has spent with the postal service is

| | |
|---|---|
| A. | 7 ± 2. |
| B. | 7 ± 1.984. |
| C. | 7 ± 0.525. |
| D. | 7 ± 0.4. |
| E. | 7 ± 0.2. |

10. Do students tend to improve their SAT Mathematics (SAT-M) score the second time they take the test? A random sample of four students who took the test twice earned the following scores.

| **Student** | 1 | 2 | 3 | 4 |
|---|---|---|---|---|
| **First Score** | 450 | 520 | 720 | 600 |
| **Second Score** | 440 | 600 | 720 | 630 |

Assume that the change in SAT-M score (second score − first score) for the population of all students taking the test twice is approximately Normally distributed with mean $\mu$. A 90% confidence interval for $\mu$ is

| | |
|---|---|
| A. | 25.0 ± 118.03. |
| B. | 25.0 ± 64.29. |
| C. | 25.0 ± 47.56. |
| D. | 25.0 ± 43.08. |
| E. | 25.0 ± 33.24. |

## Multiple Choice Answers

| Problem | Answer | Concept | Right | Wrong | Simple Mistake? | Need to Study More |
|---|---|---|---|---|---|---|
| 1 | A | Interpreting Confidence | | | | |
| 2 | C | Interpret a Confidence Interval | | | | |
| 3 | A | Width of a Confidence Interval | | | | |
| 4 | E | Biased Samples | | | | |
| 5 | C | Standard Error of $\hat{p}$ | | | | |
| 6 | C | Standard Error of $\overline{x}$ | | | | |
| 7 | C | Confidence Interval for *p* | | | | |
| 8 | B | Choosing Sample Size | | | | |
| 9 | D | Confidence Interval for $\mu$ | | | | |
| 10 | C | Confidence Interval for $\mu$ (paired data) | | | | |

## FRAPPY! Free Response AP Problem, Yay!

The following problem is modeled after actual Advanced Placement Statistics free response questions. Your task is to generate a complete, concise response in 15 minutes. After you generate your response, view two example solutions and determine whether you feel they are "complete," "substantial," "developing" or "minimal." If they are not "complete," what would you suggest to the student who wrote them to increase their score? Finally, you will be provided with a rubric. Score your response and note what, if anything, you would do differently to increase your own score.

A machine at a soft-drink bottling factory is calibrated to dispense 12 ounces of cola into cans. A simple random sample of 35 cans is pulled from the line after being filled and the contents are measured. The mean content of the 35 cans is 11.92 ounces with a standard deviation of 0.085 ounce.

a) Construct and interpret a 95% confidence interval to estimate the true mean contents of the cans being filled by this machine.

b) Based on your result from a), does the machine appear to be working properly? Justify your answer.

c) Interpret the confidence level of 95 percent in context.

**Student Response 1:**

a) One samp-t-int = (11.89, 11.94)

b) There is a 95% chance the true mean of the amount the machine fills cans is captured in this interval.

c) If we took 100 samples, 95 of them would create an interval that captures the true mean.

How would you score this response? Is it substantial? Complete? Developing? Minimal? Is there anything this student could do to earn a better score?

**Student Response 2:**

a) Conditions: Random sample is given. The cans are independent of each other. Since 35>30, we can assume normality of the sampling distribution.

   95% t-interval for the true mean contents:
   $11.92 \pm 2.042(0.085/\sqrt{(35)}) = (11.89, 11.94)$

b) We are 95% confident the true mean contents of the cans filled by this machine falls between 11.89 and 11.94 oz. It appears the machine might be underfilling the cans since 12 oz is not in the interval.

c) If we were to take many samples of size 35 and construct intervals from their sample mean contents, 95% of the intervals would capture the true mean contents being dispensed by the filling machine.

How would you score this response? Is it substantial? Complete? Developing? Minimal? Is there anything this student could do to earn a better score?

## Scoring Rubric

Use the following rubric to score your response. Each part receives a score of "Essentially Correct," "Partially Correct," or "Incorrect." When you have scored your response, reflect on your understanding of the concepts addressed in this problem. If necessary, note what you would do differently on future questions like this to increase your score.

## Intent of the Question

The goal of this question is to determine your ability to construct and interpret a confidence interval and correctly interpret the confidence level in the context of a problem.

## Solution

(a) Conditions: Random – The cans were randomly selected.
Independent – There are more than 10(35) cans on the line.
Normal – $n$ = 35 (greater than 30), so the sampling distribution of $\bar{x}$ will be approximately normal.

95% CI for $\mu$: $11.92 \pm 2.042(0.085/\sqrt{35}) = (11.89, 11.94)$

(b) We are 95% confident that the interval from 11.89 ounces to 11.94 ounces captures the true mean contents of the cans filled by this machine. It appears the machine may be filling less than it is supposed to since 12 is not in the interval.

(c) 95% of intervals constructed from random samples of 35 cans from this machine will be successful in capturing the true mean contents.

## Scoring

Parts (a), (b), and (c) are scored as essentially correct (E), partially correct (P), or incorrect (I).

**Part (a)** is essentially correct if the response correctly checks the conditions for a one-sample *t* confidence interval for a mean AND correctly calculates the interval. Part (a) is partially correct if the conditions are not properly checked but the interval is correct. Note: the construction of a *z*-interval receives a partial at most.

**Part (b)** is essentially correct if the response correctly interprets the confidence interval in context AND correctly notes the machine appears to be underfilling because 12 is not contained in the interval. Part (b) is partially correct if the interpretation lacks context OR fails to make a decision about the machine based on the interval.

**Part (c)** is essentially correct if the response correctly interprets the confidence level in context. Part (c) is partially correct if the interpretation lacks context.

4 **Complete Response**
All three parts essentially correct

3 **Substantial Response**
Two parts essentially correct and one part partially correct

2 **Developing Response**
Two parts essentially correct and no parts partially correct
One part essentially correct and two parts partially correct
Three parts partially correct

1 **Minimal Response**
One part essentially correct and one part partially correct
One part essentially correct and no parts partially correct
No parts essentially correct and two parts partially correct

# Chapter 8: Estimating with Confidence

**Across**

4. ____ t procedures allow us to compare the responses to two treatments in a matched pairs design
8. a confidence interval consists of an estimate ± margin of ____
9. to find the standard error of the sample mean, divide the sample standard deviation by the ____ of the sample size (two-words)
10. to estimate with confidence, our estimate should be calculated from a ___ sample
11. methods for drawing conclusions about a population from sample data
12. a single value used to estimate a parameter is a ____ estimator
13. we can construct a narrow interval by ____ our confidence
14. as degrees of freedom increase, the t distribution approaches the ____ distribution
16. the spread of the t distributions is ____ than the spread of the standard Normal distribution

**Down**

1. inference procedures that remain fairly accu even when a condition is violated
2. another condition for confidence intervals is observations should be ____
3. particular t distributions are specified by deg of ____
5. we can construct a narrow confidence interv by ____ our sample size
6. the margin of error consists of a ____ value the standard error of the sampling distributic
7. a ____ interval provides an estimate for a population parameter
15. confidence ____: the success rate of the method in repeated sampling

# Chapter 9: Testing A Claim

*"A statistical analysis, properly conducted, is a delicate dissection of uncertainties, a surgery of suppositions." M.J. Moroney*

## Chapter Overview

In the last chapter, you learned how to estimate a parameter based on a sample statistic. In this chapter, you will learn how to use data from a random sample to test a claim about a population parameter. Significance tests are the second type of inference you will be introduced to in this course. By the end of this chapter, you should understand the logic behind a significance test, as well as how to use a four-step procedure to carry one out. Specific methods for testing a claim about a proportion and a mean will be explored. In the next few chapters, you'll learn how to test claims about the difference between proportions or means, as well as how to perform significance tests for slopes of regression lines and distributions of categorical variables. In each case, you will use the same logic and the same four-step procedure. Be sure to take the time in this chapter to get a solid understanding of both!

## Sections in this Chapter

**Section 9.1**: Significance Tests: The Basics
**Section 9.2**: Tests About a Population Proportion
**Section 9.3**: Tests About a Population Mean

## Plan Your Learning

Use the following *suggested* guide to help plan your reading and assignments. Note: your teacher may schedule a different pacing. Be sure to follow his or her instructions!

| **Read** | 9.1: pp 527-537 | 9.1: pp 537-545 | 9.2: pp 549-555 | 9.2: pp 556-562 |
|---|---|---|---|---|
| **Do** | 1, 3, 5, 7, 9, 11, 13 | 15, 19, 21, 23, 25 | 27-30, 41, 43, 45 | 47, 49, 51, 53, 55 |

| **Read** | 9.3: pp 565-576 | 9.3: pp 577-587 | Chapter Summary |
|---|---|---|---|
| **Do** | 57-60, 71, 73 | 75, 77, 89, 94-97, 99-104 | Multiple Choice FRAPPY! |

## Section 9.1: Significance Tests: The Basics

### Before You Read: Section Summary

In this section, you will learn the basic ideas and logic behind a significance test. You will be introduced to stating hypotheses, checking conditions, calculating a test statistic, and drawing a conclusion. *P*-values will be introduced as a means of weighing the strength of evidence against a claim. Since we are basing our conclusion on a sample statistic that would vary from sample to sample, we must be aware of the fact that our conclusion could be wrong. This section ends with a discussion of the types of errors that could occur when performing a significance test and how to deal with them. Be sure to get a good understanding of the logic and format for a significance test. You will be using them throughout the rest of the course!

### "Where Am I Going?"

**Learning Targets:**

- ____ I can state correct hypotheses for a significance test about a population proportion or mean.
- ____ I can interpret *P*-values in context.
- ____ I can interpret a Type I error and a Type II error in context, and give the consequences of each.
- ____ I can describe the relationship between the significance level of a test, *P*(Type II error), and power.

### While You Read: Key Vocabulary and Concepts

significance test:

null hypothesis:

alternative hypothesis:

one-sided:

two-sided:

*P*-value:

reject $H_0$:

fail to reject $H_0$:

statistically significant:

Type I error:

Type II error:

power:

**After You Read: "Where Am I Now?"**
**Check for Understanding**

***Concept 1: The Reasoning Behind Significance Tests***

A significance test is a procedure that allows us to test a claim about a population parameter by studying a statistic from a random sample. If an observed statistic is "far" away from a hypothesized claim about a parameter, we have some evidence that the claim is wrong. Whether or not the statistic is "far enough" away depends on the sampling distribution of the statistics. That is, statistics will vary from sample to sample. What a significance test does is answers the question, "What are the chances we would observe a sample statistic at least this extreme, assuming the claim about the parameter was true?" If the chances are low, there is evidence that the claim may be wrong. If the chances are pretty good, then there is little evidence to suggest the claim is wrong. Of course, to even begin making that argument, you must start by setting up a null hypothesis (or claim about the parameter you are trying to find evidence against) and an alternative hypothesis (the claim about the parameter you are trying to find evidence for). Note: the null hypothesis will always be a statement of equality. That is, it says "Let's assume the parameter is equal to _____." The alternative hypothesis is set up to test whether or not the actual value of the parameter is greater than, less than, or simply not equal to the value in the null hypothesis.

**Check for Understanding:** ____ *I can state correct hypotheses for a significance test about a population proportion or mean.*

Suppose you suspect a "chute" of playing cards is not fair. The chute supposedly contains 10 standard decks shuffled together. You are interested in knowing whether there are more hearts than usual. To test this, you deal 12 cards at random and calculate the proportion of hearts in your hand.

a) Describe the parameter of interest in this setting.

b) Write the appropriate null and alternative hypotheses for this situation.

c) Suppose your deal contains 7 hearts. What is the sample proportion of hearts? Is it possible to deal 7 hearts out of 12 cards if the chute really does contain standard decks? Is it likely? Why or why not?

### *Concept 2: Statistical Significance*

To conclude that we have convincing evidence against a null hypothesis, our sample statistic must have a low likelihood of occurring under the assumption that the null hypothesis is true. To determine this, we must consider the sampling distribution of the statistic under the assumption that the null hypothesis is true and calculate the probability of observing a statistic at least as extreme as the one observed. If the conditions for inference are met, we can describe the sampling distribution and calculate a test statistic. This test statistic can then be used to determine a *P*-value. If this *P*-value is low enough, we can say the data are statistically significant and we have evidence to reject the null hypothesis. The general rule of thumb is to use a significance level of 5%, although some situations may specify levels of 1% or even 10%. If the *P*-value is smaller than the chosen significance level, we have enough evidence to reject the null hypothesis and conclude that the alternative is true. The reasoning behind this is that if we observe an outcome that has an extremely low chance of occurring under a given assumption, then there is evidence that the assumption may be false. If we observe an outcome under that has a fairly likely chance of occurring under a given assumption, then we have little reason to doubt the assumption!

**Check for Understanding:** ____ *I can interpret* P*-values in context.*

Refer to the hypotheses you set in the previous check for understanding. Suppose you conduct a significance test for the proportion of hearts in the chute and find the *P*-value for obtaining 7 hearts in 12 randomly selected cards is 0.004.

a) Explain what it means for the null hypothesis to be true in this setting.

b) Interpret the *P*-value in context.

c) Do these data provide convincing evidence against the null hypothesis at a 5% significance level? Explain.

### *Concept 3: Type I and Type II Errors*

Since our conclusion is based on what we would expect to see in a sampling distribution, there is a chance our sample statistic does not accurately reflect the true value of the parameter. That is, due to sampling variability, it is possible that the random sample that is collected will lead to an incorrect conclusion about the parameter. If we obtain a very small *P*-value and reject the null when, in fact, the null hypothesis is true, we commit a Type I error. The probability of committing a Type I error is equal to the significance level, $\alpha$. If we obtain a relatively large *P*-value and fail to reject the null when, in fact, the null hypothesis is false, we commit a Type II error. The power of a significance test is the probability that it will correctly reject the null hypothesis when a given alternative value of the parameter is true. You will not be expected to calculate the probability of Type II errors. However, you should be able to interpret both Type I and Type II errors in the context of the situation and explain the consequences of each.

**Understanding:** ____ *I can interpret a Type I error and a Type II error in context, and give the consequences of each and ____ I can describe the relationship between the significance level of a test, P(Type II error), and power.*

Refer to the previous checks for understanding.

a) Describe a Type I error in this setting.

b) Describe a Type II error in this setting.

## Section 9.2: Tests About a Population Proportion

### Before You Read: Section Summary

In this section, you will apply the logic of significance tests and the four-step process to test a claim about a population proportion. You will learn how to check the conditions for a significance test as well as how to calculate a test-statistic and *P*-value. Finally, you will learn how to write an appropriate conclusion in context. Your study will also include an introduction to two-sided tests to find evidence that a population proportion is different than a hypothesized value. You will then wrap up the section by studying the connection between confidence intervals and two-sided tests. Again, this section introduces you to concepts that will be used throughout the rest of the course. Make sure to get additional help if you struggle at all with the ideas behind significance testing!

### "Where Am I Going?"

**Learning Targets:**

____ I can check conditions for carrying out a test about a population proportion.
____ I can conduct a significance test about a population proportion.
____ I can use a confidence interval to draw a conclusion for a two-sided test about a population proportion.

### While You Read: Key Vocabulary and Concepts

test statistic:

significance tests – four step process:

one-sample *z* test for a proportion:

### After You Read: "Where Am I Now?" Check for Understanding

### *Concept 1: One-Sample z Test for a Proportion*

Significance tests and confidence intervals are both based on the same concept—sampling distributions. Therefore, when performing inference with either of these methods, it is important to follow a process that ensures the calculations are justified. To test a claim about a population parameter, the following four-step process should be used:

1) **STATE the parameter of interest and the hypotheses you would like to test.**
   When testing a claim about a population proportion, we start by defining hypotheses about the parameter. The null hypothesis assumes that the population proportion is equal to a particular value while the alternative hypothesis is that the population proportion is greater than, less than, or not equal to that value:

$$H_0: p = p_0$$
$$H_a: p > p_0 \quad \text{OR} \quad p < p_0 \quad \text{OR} \quad p \neq p_0$$

   State a significance level.

2) **PLAN: Choose the appropriate inference method and check the conditions.**
   We must check the Random, Normal, and Independent conditions before proceeding with inference. To ensure that the sampling distribution of $\hat{p}$ is approximately Normal, check to make sure $np_0$ and $n(1 - p_0)$ are both at least 10. If sampling is done without replacement, verify that the population is at least 10 times as large as the sample.

3) **DO: If conditions are met, calculate a test statistic and *P*-value.**
   Compute the $z$ test statistic
   $$z = \frac{\hat{p} - p_0}{\sqrt{\dfrac{p_0(1 - p_0)}{n}}}$$
   and find the *P*-value by calculating the probability of observing a $z$ statistic at least this extreme in the direction of the alternative hypothesis.

4) **CONCLUDE by interpreting the results of your calculations in the context of the problem.**
   If the *P*-value is smaller than the stated significance level, you have significant evidence to reject the null hypothesis. If it is greater than or equal to the significance level, then you fail to reject the null hypothesis.

This four-step procedure can be used with *any* significance test! The only things that will change are the conditions that must be checked and the test statistic calculation. Be sure to get a good grasp of this process here as you will be using it a lot in the coming chapters!

**Check for Understanding:** ____ *I can check conditions for carrying out a test about a population proportion and* ____ *I can conduct a significance test about a population proportion.*

A study of classic authors uncovered a distinguishable speech pattern that differed from author to author. Plato utilized this pattern in 21.4% of the passages in his works. The owner of a rare bookstore claims to have an original Plato work, but you suspect the speech pattern occurs too frequently to be an original Plato work. A random sample of passages from the work in question was taken and it was found that 136 of the 439 selected passages followed the speech pattern. Do these data provide convincing evidence that the work was not written by Plato? Conduct a one-sample $z$ test using the four-step procedure.

### *Concept 2: Confidence Intervals Give More Information*

Significance tests provide evidence that supports or rejects an assumption about a population parameter. If an observed sample statistic is far enough away from a parameter's assumed value, a significance test will suggest there is evidence to question that assumed value. However, the test does not say anything about what the actual parameter value may be. A confidence interval provides more information by suggesting a range of plausible values for the parameter.

**Check for Understanding:** ____ *I can use a confidence interval to draw a conclusion for a two-sided test about a population proportion.*

A recent study suggested that 77% of teenagers have texted while driving. A random sample of 27 teenage drivers in Atlanta was taken and 15 admitted to texting while driving. Use a 99% confidence interval to determine whether there is convincing evidence that the population proportion of teens who text while driving is different than 77%.

## Section 9.3: Tests About a Population Mean

### Before You Read: Section Summary

In this section, you'll learn how to perform a significance test about a population mean μ. Just like you did for proportions, you'll learn how to set up hypotheses, check the appropriate conditions, calculate a test statistic and *P*-value, and draw a conclusion in context. You will also learn how tests involving "paired data" can be performed using one-sample *t* procedures.

### "Where Am I Going?"

**Learning Targets:**

____ I can check conditions for carrying out a test about a population mean.
____ I can conduct a one-sample *t* test about a population mean $\mu$.
____ I can use a confidence interval to draw a conclusion for a two-sided test about a population mean.
____ I can recognize paired data and use one-sample *t* procedures to perform significance tests for such data.

### While You Read: Key Vocabulary and Concepts

one-sample *t* test:

paired data:

paired *t* procedures:

### After You Read: "Where Am I Now?" Check for Understanding

### *Concept 1: One Sample t Test for μ*

The process for testing a claim about a population mean follows the same format as the process used for a population proportion. However, like confidence intervals for means, you usually need to base your calculations on the *t* distributions (unless $\sigma$ is somehow known).

1) **STATE the parameter of interest and the hypotheses you would like to test.**
   The null hypothesis assumes that the population mean is equal to a particular value while the alternative hypothesis is that the population mean is greater than, less than, or not equal to that value:
   $H_0$: $\mu = \mu_0$
   $H_a$: $\mu > \mu_0$ OR $\mu < \mu_0$ OR $\mu \neq \mu_0$
   State a significance level.

2) **PLAN: Choose the appropriate inference method and check the conditions.**
   We must check the Random, Normal, and Independent conditions before proceeding with inference. To ensure that the sampling distribution of $\bar{x}$ is approximately Normal, check that the population distribution is Normal OR $n \geq 30$. If $n < 30$, it is sufficient if the sample data show no signs of strong skewness or outliers. If sampling is done without replacement, verify that the population is at least 10 times as large as the sample.

3) **DO: If conditions are met, calculate a test statistic and *P*-value.**

Compute the *t* test statistic

$$t = \frac{\bar{x} - \mu_0}{\frac{s_x}{\sqrt{n}}}$$

and find the *P*-value by calculating the probability of observing a *t* statistic at least this extreme in the direction of the alternative hypothesis in a *t* distribution with $n - 1$ degrees of freedom.

4) **CONCLUDE by interpreting the results of your calculations in the context of the problem.**

If the *P*-value is smaller than the stated significance level, you have significant evidence to reject the null hypothesis. If it is larger than the significance level, then you fail to reject the null hypothesis.

As with proportions, a confidence interval for a population mean can be used to test a two-sided claim. Further, constructing a confidence interval gives a range of plausible values for $\mu$, while a significance test only allows you to conclude that $\mu$ may be different from a particular value.

**Check for Understanding:** ____ *I can check conditions for carrying out a test about a population mean and* ____ *I can conduct a one-sample* t *test about a population mean* μ.

Humerus bones from the same species of animal tend to have approximately the same length-to-width ratios. When fossils of humerus bones are discovered, archeologists can often determine the species of animal by examining these ratios. It is known that the species Molekius Primatium exhibits a mean ratio of $\mu = 8.9$. Suppose 41 fossils of humerus bones are unearthed at a site on Minnesota's Iron Range, where this species was known to have lived. Researchers are willing to view these as a random sample of all such humerus bones. The length-to-width ratios were calculated and are listed below. Test whether the population mean for the species that left these bones differs from 8.9 at $\alpha$=.05.

| | | | | | | | | | | | |
|---|---|---|---|---|---|---|---|---|---|---|---|
| 9.73 | 10.89 | 9.07 | 9.2 | 9.33 | 9.98 | 9.84 | 9.59 | 9.48 | 8.71 | 9.57 | 9.29 |
| 9.94 | 8.07 | 8.37 | 6.85 | 8.52 | 8.87 | 6.23 | 9.41 | 6.66 | 9.35 | 8.86 | 9.93 |
| 8.91 | 11.77 | 10.48 | 10.39 | 9.39 | 9.17 | 9.89 | 8.17 | 8.39 | 8.8 | 10.02 | 8.38 |
| 11.67 | 8.3 | 9.17 | 12 | 9.38 | | | | | | | |

### *Concept 2: Paired Data*

Studies that involve making two observations on the same individual or making an observation on each of two very similar individuals result in paired data. In these types of studies, we are often interested in analyzing the differences in responses within each pair. If the conditions for inference are met, we can use one-sample *t* procedures to estimate or test a claim about the mean difference $\mu_d$. In this case, we refer to the inference method as a paired *t* procedure.

**Check for Understanding:** ____ *I can construct and interpret a confidence interval for a population mean.*

A study measured how fast subjects could repeatedly push a button when under the effects of caffeine. Subjects were asked to push a button as many times as possible in two minutes after consuming a typical amount of caffeine. During another test session, they were asked to push the button after taking a placebo. The subjects did not know which treatment they were administered each day and the order of the treatments was randomly assigned. The data, given in presses per two minutes for each treatment follows. Use a paired *t* procedure to determine whether or not caffeine results in a higher rate of beats, on average, per two-minute period.

| Subject | Beats Caffeine | Beats Placebo | |
|---|---|---|---|
| 1 | 281 | 201 | |
| 2 | 284 | 262 | |
| 3 | 300 | 283 | |
| 4 | 421 | 290 | |
| 5 | 240 | 259 | |
| 6 | 294 | 291 | |
| 7 | 377 | 354 | |
| 8 | 345 | 346 | |
| 9 | 303 | 283 | |
| 10 | 340 | 391 | |
| 11 | 408 | 411 | |

## Chapter Summary: Testing a Claim

A significance test tells us whether or not a sample provides convincing evidence against a claim about a population parameter. The test answers the question, "How likely would it have been to observe this particular sample statistic (or one more extreme) if the claim about the parameter was true?" This probability, the *P*-value, gives us an idea just how surprising an observed statistic is under the assumption that the null hypothesis is true. If the *P*-value is small (less than 5% or some other chosen significance level), we have enough evidence to reject the null hypothesis, suggesting the actual parameter value may be greater than, less than, or different from the claim. If the *P*-value is greater than or equal to a specified significance level, we do not have convincing evidence against the claim about the parameter. Keep in mind, however, that since we are basing our decision on a likelihood from a sampling distribution, it is possible that we may be wrong. Type I and Type II errors occur when we mistakenly reject a null hypothesis that is true or fail to reject a null hypothesis that is false, respectively. You should be able to define each of these errors in the context of the situation as well as explain the consequences of each.

This chapter not only introduced you to the logic behind significance tests, but also a four-step procedure for carrying them out. This procedure will be used throughout the rest of the course, so hopefully you are comfortable with it at this point! If not, be sure to practice a few more tests and focus on stating the parameter and hypotheses, checking the appropriate conditions, calculating a test statistic and *P*-value, and concluding in the context of the problem. If you follow those steps, you should have very few problems performing significance tests in the upcoming chapters!

### After You Read: "How Can I Close the Gap?"

Complete the vocabulary puzzle, multiple choice questions, and FRAPPY. Check your answers and on your performance on each of the targets.

| Target | Got It! | Almost There | Needs Some Work |
|---|---|---|---|
| I can state correct hypotheses for a significance test about a population proportion or mean. | | | |
| I can interpret *P*-values in context. | | | |
| I can interpret a Type I error and a Type II error in context, and give the consequences of each. | | | |
| I can check conditions for carrying out a test about a population proportion. | | | |
| I can conduct a significance test about a population proportion. | | | |
| I can use a confidence interval to draw a conclusion for a two-sided test about a population proportion. | | | |
| I can check conditions for carrying out a test about a population mean. | | | |
| I can conduct a one-sample $t$ test about a population mean $\mu$. | | | |
| I can use a confidence interval to draw a conclusion for a two-sided test about a population mean. | | | |
| I can recognize paired data and use one-sample $t$ procedures to perform significance tests for such data. | | | |

Did you check "Needs Some Work" for any of the targets? If so, what will you do to address your needs for those targets?

*Learning Plan:*

| |
|---|
| |

# Chapter 9 Multiple Choice Practice

**Directions.** *Identify the choice that best completes the statement or answers the question. Check your answers and note your performance when you are finished.*

1. The average yield of a certain crop is 10.1 bushels per plant. A biologist claims that a new fertilizer will result in a greater yield when applied to the crop. A random sample of 25 of plants given the fertilizer has an average yield of 10.8 bushels and a standard deviation of 2.1 bushels. The appropriate null and alternative hypotheses to test the biologist's claim are

| | |
|---|---|
| A. | $H_0$: $\mu = 10.8$ against $H_a$: $\mu > 10.8$. |
| B. | $H_0$: $\mu = 10.8$ against $H_a$: $\mu \neq 10.8$. |
| C. | $H_0$: $\mu = 10.1$ against $H_a$: $\mu > 10.1$. |
| D. | $H_0$: $\mu = 10.1$ against $H_a$: $\mu < 10.1$. |
| E. | $H_0$: $\mu = 10.1$ against $H_a$: $\mu \neq 10.1$. |

2. An opinion poll asks a random sample of 200 adults how they feel about voting for an amendment in an upcoming election. In all, 150 say they are in favor of the amendment. Does the poll provide evidence that the proportion *p* of adults who are in favor of the amendment is greater than 60%? The null and alternative hypotheses are

| | |
|---|---|
| A. | $H_0$: $p = 0.6$ against $H_a$: $p > 0.6$. |
| B. | $H_0$: $p = 0.6$ against $H_a$: $p \neq 0.6$. |
| C. | $H_0$: $p = 0.6$ against $H_a$: $p < 0.6$. |
| D. | $H_0$: $p = 0.6$ against $H_a$: $p = 0.75$. |
| E. | $H_0$: $p = 0.75$ against $H_a$: $p < 0.6$. |

3. A test of significance produces a *P*-value of 0.035. Which of the following conclusions is appropriate?

| | |
|---|---|
| A. | Accept $H_a$ at the $\alpha = 0.05$ level |
| B. | Reject $H_a$ at the $\alpha = 0.01$ level |
| C. | Fail to reject $H_0$ at the $\alpha = .05$ level |
| D. | Reject $H_0$ at the $\alpha = 0.05$ level |
| E. | Accept $H_0$ at the $\alpha = 0.01$ level |

4. A Type II error is

| | |
|---|---|
| A. | rejecting the null hypothesis when it is true. |
| B. | failing to reject the null hypothesis when it is false. |
| C. | rejecting the null hypothesis when it is false. |
| D. | failing to reject the null hypothesis when it is true. |
| E. | more serious than a Type I error. |

5. A researcher plans to conduct a significance test at the $\alpha = 0.05$ significance level. She designs her study to have a power of 0.85 for a particular alternative value of the parameter. The probability that the researcher will commit a Type II error for the particular alternative value of the parameter at which she computed the power is

| | |
|---|---|
| A. | 0.05. |
| B. | 0.15. |
| C. | 0.80. |
| D. | 0.95. |
| E. | equal to the 1 - (*P*-value) and cannot be determined until the data have been collected. |

6. In hypothesis testing $\beta$ is the probability of committing a Type II error in a test with significance level $\alpha$. The probability of committing a Type I error is

| | |
|---|---|
| A. | $1 - \beta$ |
| B. | $1 - \alpha$ |
| C. | $\beta - \alpha$ |
| D. | $\alpha$ |
| E. | Can not be determined |

7. A claimed psychic was presented with 200 cards face down and asked to determine if the card was one of five symbols: a star, cross, circle, square, or three wavy lines. The "psychic" was correct in 50 cases. To determine if he has ESP, we test the hypotheses $H_0$: $p = 0.20$, $H_a$: $p > 0.20$, where $p$ represents the true proportion of cards for which the psychic would correctly identify the symbol in the long run. Assume the conditions for inference are met. The *P*-value of this test is

| | |
|---|---|
| A. | between .10 and .05. |
| B. | between .05 and .025. |
| C. | between .025 and .01. |
| D. | between .01 and .001. |
| E. | below .001. |

8. The most important condition for drawing sound conclusions from statistical inference is usually

| | |
|---|---|
| A. | that the population standard deviation is known. |
| B. | that at least 30 people are included in the study. |
| C. | that the data come from a random sample or a randomized experiment. |
| D. | that the population distribution is exactly Normal. |
| E. | that no calculation errors are made in the confidence interval or test statistic. |

9. The mean weight of a random sample of 35 athletes is found to be 165 pounds with a standard deviation of 20 pounds. It is believed that a mean weight of 160 pounds would be normal for this group. To see if there is evidence that the mean weight of the population of all athletes of this type is significantly higher than 160 pounds, the hypotheses $H_0$: $\mu = 160$ vs. $H_a$: $\mu > 160$ are tested. You obtain a *P*-value of 0.0742. Which of the following is true?

| | |
|---|---|
| A. | At the 5% significance level, you have proved that $H_0$ is true. |
| B. | You have failed to obtain sufficient evidence against $H_0$. |
| C. | At the 5% significance level, you have failed to prove that $H_0$ is true, and a larger sample size is needed to do so. |
| D. | Only 7.42% of the athletes weigh less than 160 pounds. |
| E. | None of the above. A significance test is inappropriate in this setting. |

10. A medical researcher wishes to investigate the effectiveness of exercise versus diet in losing weight. Two groups of 25 overweight adult subjects are used, with a subject in each group matched to a similar subject in the other group on the basis of a number of physiological variables. One of the groups is placed on a regular program of vigorous exercise but with no restriction on diet, and the other is placed on a strict diet but with no requirement to exercise. The weight losses after 20 weeks are determined for each subject, and the difference between matched pairs of subjects (weight loss of subject in exercise group - weight loss of matched subject in diet group) is computed. The mean of these differences in weight loss is found to be 2 lb with standard deviation $s_x = 4$ lb. Is this convincing evidence of a difference in mean weight loss for the two methods? To answer this question, you should use

| | |
|---|---|
| A. | one-proportion *z* test. |
| B. | one-sample *z* interval for $\mu_d$. |
| C. | one-proportion *z* interval. |
| D. | one-sample *t* test for $\mu_d$. |
| E. | none of the above. |

## Multiple Choice Answers

| Problem | Answer | Concept | Right | Wrong | Simple Mistake? | Need to Study More |
|---|---|---|---|---|---|---|
| 1 | C | Writing Hypotheses | | | | |
| 2 | A | Writing Hypotheses | | | | |
| 3 | D | Significance Level and *P*-value | | | | |
| 4 | B | Type I and Type II Errors | | | | |
| 5 | B | Power | | | | |
| 6 | D | Type I and Type II Errors | | | | |
| 7 | B | Calculating *P*-value for proportions | | | | |
| 8 | C | Conditions for Inference | | | | |
| 9 | B | Interpreting *P*-value | | | | |
| 10 | D | Paired *t*-test | | | | |

## FRAPPY! Free Response AP Problem, Yay!

The following problem is modeled after actual Advanced Placement Statistics free response questions. Your task is to generate a complete, concise response in 15 minutes. After you generate your response, view two example solutions and determine whether you feel they are "complete," "substantial," "developing" or "minimal." If they are not "complete," what would you suggest to the student who wrote them to increase their score? Finally, you will be provided with a rubric. Score your response and note what, if anything, you would do differently to increase your own score.

During a recent movie promotion, Fruity O's cereal placed mini action figures in some of its boxes. The advertisement on the box states 1 out of every 4 boxes contains an action figure. A group of promotional-toy collectors suspects the proportion of boxes containing the action figure may be lower than 0.25. The group purchased 70 boxes of cereal and found 12 action figures. Assuming the 70 boxes represent a random sample of all of the cereal boxes, is there evidence to support the toy collector's belief that the proportion of boxes containing the figure is less than 0.25? Provide statistical evidence to support your answer.

**Student Response 1:**

We will perform a one-sample z-test for the proportion of boxes, *p*, that contain the action figures.

$H_0$: $p = 0.25$
$H_a$: $p < 0.25$

We are told the sample is random. Further, both *np* and *n*(1 – *p*) are greater than 10.

$z = \dfrac{0.17 - 0.25}{\sqrt{\dfrac{.25(.75)}{70}}} = -1.52$ and our *p*-value is 0.064.

We do not have significant evidence at the 5% level to reject the null hypothesis. The proportion of boxes that contain the action figure is 0.25.

How would you score this response? Is it substantial? Complete? Developing? Minimal? Is there anything this student could do to earn a better score?

**Student Response 2:**

Let *p* = the actual proportion of cereal boxes that contain the action figure

$H_0$: $p = 0.25$
$H_a$: $p < 0.25$

We are told the sample is random.
70(0.25) = 17.5 and 70(0.75) = 52.5 Both are greater than 10.
There are more than 700 cereal boxes in the population.

One sample z test for a proportion

$z = z = \dfrac{0.17 - 0.25}{\sqrt{\dfrac{.25(.75)}{70}}} = -1.52$ *p*-value = 0.064.

Since our p-value is greater than the typical significance level of 5%, we do not have significant evidence to reject the null hypothesis. There is not enough support to suggest that the actual proportion of boxes that contain the figure is less than 0.25.

How would you score this response? Is it substantial? Complete? Developing? Minimal? Is there anything this student could do to earn a better score?

## Scoring Rubric

Use the following rubric to score your response. Each part receives a score of "Essentially Correct," "Partially Correct," or "Incorrect." When you have scored your response, reflect on your understanding of the concepts addressed in this problem. If necessary, note what you would do differently on future questions like this to increase your score.

### Intent of the Question

The goal of this question is to determine your ability to conduct a significance test for a single proportion.

### Solution

The solution should contain 4 parts:

- Hypotheses must be stated appropriately. $H_0$: $p = 0.25$ and $H_a$: $p < 0.25$ where $p$ = true proportion of boxes containing the action figure.
- Name of test and conditions: The test must be identified by name or formula as a one-sample z test for a population proportion. **Random** sample is given. **Normal:** 70(0.25) = 17.5 and 70(0.75) = 52.5 Both are at least 10. **Independent:** There are more than 700 cereal boxes in the population.
- Mechanics: $z = -1.52$ and *P*-value = 0.064
- Conclusion: Because the *P*-value is greater than the typical significance level of 5%, we fail to reject the null hypothesis. There is not sufficient evidence to suggest that the actual proportion of boxes with the action figure is less than 0.25.

### Scoring

Each element scored as essentially correct (E), partially correct (P), or incorrect (I).

**Hypotheses** is essentially correct if the hypotheses are written correctly. This part is partially correct if the test isn't identified or if the hypotheses are written incorrectly.

**Name & Conditions** is essentially correct if the response correctly identifies the test by name or formula AND correctly checks the Normal condition and 10% condition. This part is partially correct if the test is correctly identified but only one of the conditions is checked correctly OR if the test is not identified correctly but both conditions are checked correctly.

**Mechanics** is essentially correct if the response correctly calculates the test statistic and p-value. This part is partially correct if one of the calculations is incorrect.

**Conclusion** is essentially correct if the response correctly fails to reject the null hypothesis because the P-value is greater than a significance level of 5% OR if the response is to reject the null hypothesis because the P-value is less than a significance level of 10% and provides an interpretation in context. This part is partially correct if the response fails to justify the decision by comparing the P-value to a significance level OR if the conclusion lacks an interpretation in context.

### Scoring

This problem has four elements, each receiving an E, P, or I. Assign one point to each E, 0.5 points to each P, and 0 points to each I. Total the points to determine your score. If a score falls between two whole values, consider the strength of the entire response to determine whether to round up or down.

# Chapter 9: Testing a Claim

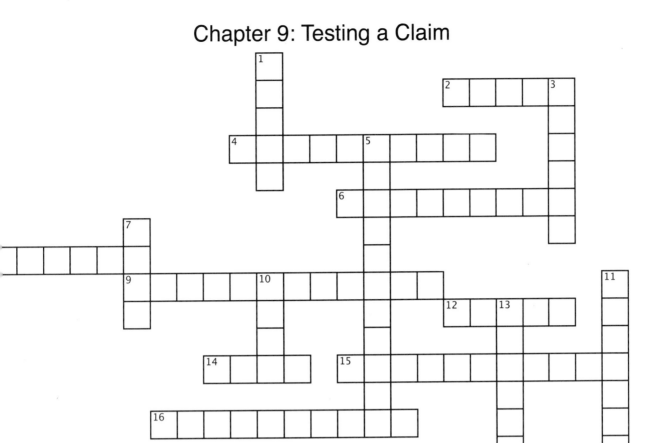

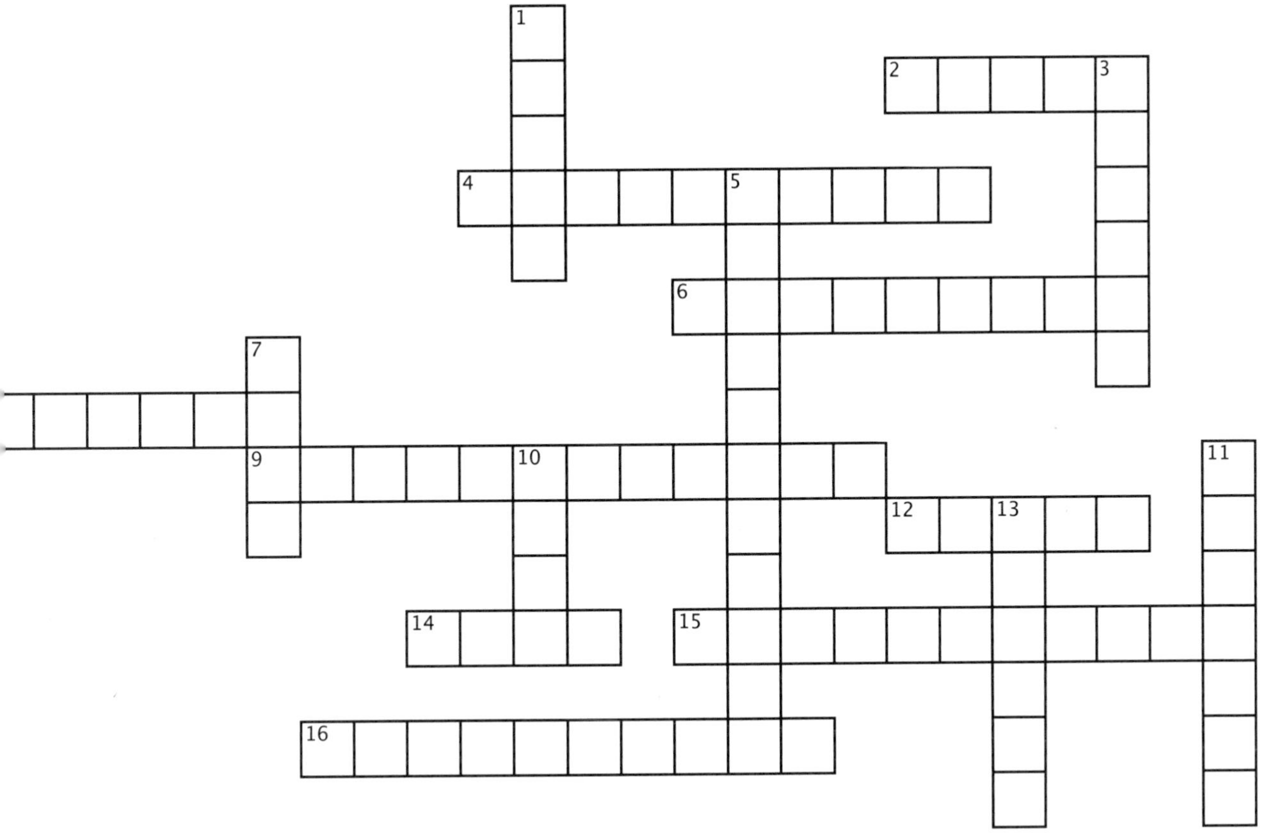

## 'oss

the probability that a significance test will reject the null when a particular alternative value of the paramter is true

hypotheses always refer to the _____

the test _____ is a standardized value that assesses how far the estimate is from the hypothesized parameter

the probability that we would observe a statistic at least as extreme as the one observed, assuming the null is true (two terms)

we can use a ______ test to compare observed data with a hypothesis about a population

greek letter used to designate the significance level

the _____ hypothesis is a the claim for which we are seeking evidence against

an observed difference that is too small to have occured due to chance alone is considered statistically _____

the statements a statistical test is designed to compare

## Down

1. if we reject the null hypothesis when it is actually true, we commit a Type I _____
3. if we calculate a very small P value, we have evidence to _____ the null
5. the _____ hypothesis is the claim about the population for which we are finding evidence for
7. reject the null hypothesis if the P value is _____ than the significance level
10. if our calculated P value is not small enough to provide convincing evidence, we ___ to reject he null
11. conclusions should always be written in _____
13. a _____ test allows us to analyze differences in responses within pairs

# Chapter 10: Comparing Two Populations or Groups

*"We must be careful not to confuse data with the abstractions we use to analyze them." William James*

## Chapter Overview

Up to this point, our studies have focused on inference for a single parameter. However, many statistical studies involve comparing two populations or treatments. In this chapter, we will expand our collection of inference procedures to include confidence intervals and significance tests for the difference between two proportions or means. The procedures follow the exact same format as those we learned for a single parameter. The only difference is that now you will need to rely on the sampling distribution of the difference between proportions or means to perform your calculations. You will learn the characteristics of each of those sampling distributions in this chapter and how to use them to construct a confidence interval or perform a significance test. As you did in the last chapter, you will follow a four-step process for carrying out the inference procedures. Your job now will be to make sure you understand *when* to use each one!

## Sections in this Chapter



## Plan Your Learning

Use the following *suggested* guide to help plan your reading and assignments. Note: your teacher may schedule a different pacing. Be sure to follow his or her instructions!

| **Read** | 10.1: pp 601-607 | 10.1: pp 608-611 | 10.1: pp 611-620 | 10.2: pp 627-632 |
|---|---|---|---|---|
| **Do** | 1, 3, 5 | 7, 9, 11, 13 | 15, 17, 21, 23 | 29-32, 35, 37, 57 |

| **Read** | 10.2: pp 633-637 | 10.3: pp 638-651 | Chapter Summary |
|---|---|---|---|
| **Do** | 39, 41, 43, 45 | 51, 53, 59, 65, 67-70 | Multiple Choice FRAPPY! |

## Section 10.1: Comparing Two Proportions

### Before You Read: Section Summary

In this section, you will learn how to compare two population proportions using confidence intervals and significance tests. To begin, you will explore the sampling distribution of a difference between two proportions. Like you did with a single proportion, you will learn the conditions necessary for performing inference and then how to construct a confidence interval or perform a significance test. By the end of this section, you should be able to provide an estimate of the difference between two proportions and test a claim about the difference between two proportions.

### "Where Am I Going?"

**Learning Targets:**

- ____ I can describe the characteristics of the sampling distribution of $\hat{p}_1 - \hat{p}_2$
- ____ I can calculate probabilities using the sampling distribution of $\hat{p}_1 - \hat{p}_2$
- ____ I can determine whether or not the conditions for performing inference are met.
- ____ I can construct and interpret a confidence interval to compare two proportions.
- ____ I can perform a significance test to compare two proportions.
- ____ I can interpret the results of inference procedures in a randomized experiment.

### While You Read: Key Vocabulary and Concepts

sampling distribution of $\hat{p}_1 - \hat{p}_2$:

standard error:

two-sample *z* interval for a difference between two proportions:

pooled (combined) sample proportion:

two-sample *z* test for the difference between two proportions:

randomization distribution:

## After You Read: "Where Am I Now?" Check for Understanding

### *Concept 1: Sampling Distribution of* $\hat{p}_1 - \hat{p}_2$

If we want to compare two population proportions based on data from independent random samples, we must consider what would happen in repeated sampling. That is, we must explore the sampling distribution of the difference between two proportions. If we take repeated samples of the same size $n_1$ from a population with proportion of interest $p_1$, and samples of size $n_2$ from a population with proportion of interest $p_2$, the sampling distribution of $\hat{p}_1 - \hat{p}_2$ will have the following characteristics:

1) The shape of the sampling distribution will be approximately Normal if $n_1p_1$, $n_1(1-p_1)$, $n_2p_2$, and $n_2(1-p_2)$ are all at least 10.
2) The mean of the sampling distribution is $p_1 - p_2$.
3) The standard deviation of the sampling distribution is $\sqrt{\frac{p_1(1-p_1)}{n_1}+\frac{p_2(1-p_2)}{n_2}}$

Note: The formula for the standard deviation of the sampling distribution is exactly correct if sampling is done with replacement or if the populations are infinite. It is approximately correct if sampling is done without replacement as long as the 10% condition is satisfied: the sample sizes must be less than or equal to 10% of the size of the populations of interest.

---

**Check for Understanding:** ____ *I can describe the characteristics of the sampling distribution of* $\hat{p}_1 - \hat{p}_2$ *and* ____ *I can calculate probabilities using the sampling distribution of* $\hat{p}_1 - \hat{p}_2$

School officials are interested in implementing a policy that would allow students to bring their own technology to school for academic use. There are two large high schools in a town, Lakeville North and Lakeville South, each with 1700 students. At Lakeville North, 60% of students own technological devices that could be used for academic purposes. 75% of students at Lakeville South own those types of devices. The district takes an SRS of 125 students from Lakeville North and a separate SRS of 160 students at Lakeville South. The sample proportions of students who own devices that could be used at the school are recorded and the difference, $\hat{p}_S - \hat{p}_N$, is determined to be 0.07.

a) Describe the shape, center, and spread of the sampling distribution of $\hat{p}_S - \hat{p}_N$.

b) Find the probability of getting a difference in sample proportions of 0.07 or less from the two surveys. Show your work.

c) Does the result in part (b) give you reason to doubt the study's reported value? Explain.

### *Concept 2: Conference Intervals for $p_1 - p_2$*

To construct a confidence interval for the difference between two population proportions $p_1 - p_2$, you should follow the four step process introduced in Chapter 8. The logic behind its construction is the same as the logic behind the construction of a confidence interval for a single proportion. That is, we will estimate the difference by comparing the sample proportions from two random samples and build an interval around that point estimate by using the standard error of the statistic and a critical *z* value. To construct a two-sample *z* interval for a difference between two proportions:

- **State** the parameters of interest and the confidence level you will be using to estimate the difference.
- **Plan:** Indicate the type of confidence interval you are constructing and verify that the conditions for Random, Normal, and Independent are satisfied for the two samples.
- **Do** the actual construction of the interval using the following formula:

$$(\hat{p}_1 - \hat{p}_2) \pm z^* \sqrt{\frac{\hat{p}_1(1-\hat{p}_1)}{n_1} + \frac{\hat{p}_2(1-\hat{p}_2)}{n_2}}$$

  where $z^*$ is the critical value for the standard Normal curve with area C between $-z^*$ and $z^*$.
- **Conclude** by interpreting the interval in the context of the problem.

**Check for Understanding:** ____ *I can determine whether or not the conditions for performing inference are met and* ____ *I can construct and interpret a confidence interval to compare two proportions.*

In 1990, 551 of 1500 randomly sampled adults indicated they smoked. In 2010, 652 of 2000 randomly sampled adults indicated they smoked. Use this information to construct and interpret a 95% confidence interval for the difference in the proportion of adults who smoke in 1990 and 2010.

### *Concept 3: Significance tests for $p_1 - p_2$*

To test a claim about the difference between two proportions, we will follow the same four-step process learned in Chapter 9. The key difference when performing a test about $p_1 - p_2$ is that when we assume these two parameters are equal in the null hypotheses, we need to estimate just what value they are equal to. To calculate this "pooled" proportion, simply divide the sum of the successes in the two samples by the sum of the sample sizes. This pooled proportion can then be used in the formula for the standard deviation of the sampling distribution. To conduct a two-sample z test for the difference between two proportions:

1) **STATE the parameter of interest and the hypotheses you would like to test.**
The null hypothesis assumes that the parameters are equal while the alternative hypothesis is that one proportion is greater than, less than, or not equal to the other:

$H_0$: $p_1 = p_2$

$H_a$: $p_1 > p_2$ OR $p_1 < p_2$ OR $p_1 \neq p_2$

State a significance level.

2) **PLAN: Choose the appropriate inference method and check the conditions.**
We must check the Random, Normal, and Independent conditions. Verify that the data come from random samples or the groups in a randomized experiment. To ensure that the sampling distribution of $\hat{p}_1 - \hat{p}_2$ is approximately Normal, check that $n_1\hat{p}_1$, $n_1(1-\hat{p}_1)$, $n_2\hat{p}_2$, and $n_2(1-\hat{p}_2)$ are all at least 10. If sampling is done without replacement, check that the populations are at least 10 times as large as the samples.

3) **DO: If conditions are met, calculate a test statistic and *P*-value.**
Compute the pooled proportion $\hat{p}_C = \dfrac{X_1 + X_2}{n_1 + n_2}$ and $z$ test statistic

$$z = \frac{(\hat{p}_1 - \hat{p}_2) - 0}{\sqrt{\dfrac{\hat{p}_C(1-\hat{p}_C)}{n_1} + \dfrac{\hat{p}_C(1-\hat{p}_C)}{n_2}}}$$

and find the *P*-value by calculating the probability of observing a $z$ statistic at least this extreme in the direction of the alternative hypothesis.

4) **CONCLUDE by interpreting the results of your calculations in the context of the problem.**
If the *P*-value is smaller than the stated significance level, you can conclude that you have significant evidence to reject the null hypothesis. If it is larger than or equal to the significance level, then you fail to reject the null hypothesis.

**Understanding:** ____ *I can perform a significance test to compare two proportions.*

A school official suspects the difference in the proportion of students who own technological devices between Lakeville North and Lakeville South high schools may be a result of a difference in the socioeconomic status of the students in the two schools. The results of a random sampling of student registration records indicated 28 out of 120 students at Lakeville North came from low-income families while 30 out of 150 students at Lakeville South came from low income families. Do these data provide convincing evidence that the proportion of low income students at Lakeville North is higher than the proportion at Lakeville South? Use a 5% significance level.

## Section 10.2: Comparing Two Means

### Before You Read: Section Summary

In this section, you will learn to compare two means instead of two proportions. You will start by exploring the sampling distribution of the difference between two means $\bar{x}_1 - \bar{x}_2$ along with the conditions necessary to perform inference. Then you will learn how to estimate a difference between two means as well as how to test a claim about that difference. Like you did with a single mean, you will rely on *t* distributions when performing calculations. Because some of the calculations are complex, you may wish to rely on your calculator to do most of the work. As always, though, you will want to make sure you can interpret the results that your calculator gives you!

### "Where Am I Going?"

**Learning Targets:**

- _____ I can describe the characteristics of the sampling distribution of $\bar{x}_1 - \bar{x}_2$
- _____ I can calculate probabilities using the sampling distribution of $\bar{x}_1 - \bar{x}_2$
- _____ I can determine whether or not the conditions for performing inference are met.
- _____ I can construct and interpret a two-sample *t* interval to compare two means.
- _____ I can perform a two-sample *t* test to compare two means.
- _____ I can interpret the results of inference procedures in a randomized experiment.
- _____ I can determine the proper inference procedure to use in a given setting.

### While You Read: Key Vocabulary and Concepts

sampling distribution of $\bar{x}_1 - \bar{x}_2$:

two-sample *t* interval for $\mu_1 - \mu_2$:

two-sample *t* test for $\mu_1 - \mu_2$:

### After You Read: "Where Am I Now?" Check for Understanding

**_Concept 1: Sampling Distribution of $\bar{x}_1 - \bar{x}_2$_**

Like proportions, if we want to compare two means based on the results of independent random samples or randomly assigned groups, we must consider what would happen in repeated randomization. That is, we must explore the sampling distribution of the difference between two means. If we take repeated samples of the same size $n_1$ from a population with mean $\mu_1$ and standard deviation $\sigma_1$ and samples of size $n_2$ from a population with mean $\mu_2$ and standard deviation $\sigma_2$, the sampling distribution of $\bar{x}_1 - \bar{x}_2$ will have the following characteristics:

1) The shape of the sampling distribution will be Normal if the population distributions are Normal OR approximately Normal if both sample sizes are at least 30.

2) The mean of the sampling distribution is $\mu_1 - \mu_2$.

3) The standard deviation of the sampling distribution is $\sqrt{\frac{\sigma_1^2}{n_1}+\frac{\sigma_2^2}{n_2}}$

Note: The formula for the standard deviation of the sampling distribution is exactly correct if sampling is done with replacement or if the populations are infinite. It is approximately correct if sampling is done without replacement as long as the 10% condition is satisfied: the sample sizes must be less than or equal to 10% of the size of the populations of interest.

**Check for Understanding:** ____ *I can describe the characteristics of the sampling distribution of* $\bar{x}_1 - \bar{x}_2$ *and* ____ *I can calculate probabilities using the sampling distribution of* $\bar{x}_1 - \bar{x}_2$.

Researchers are interested in studying the effect of sleep on exam performance. Suppose the population of individuals who get at least 8 hours of sleep prior to an exam score an average of 96 points on the exam with a standard deviation of 18 points. The population of individuals who get less than 8 hours of sleep score an average of 72 points with a standard deviation of 9.4 points. Suppose 40 individuals are randomly sampled from each population.

a) Describe the shape, center, and spread of the sampling distribution of $\bar{x}_1 - \bar{x}_2$

b) Find the probability of observing a difference in sample means of 2 points or more from the two samples. Show your work.

### *Concept 2: Conference Intervals for $\mu_1 - \mu_2$*

To construct a confidence interval for the difference between two population or treatment means $\mu_1 - \mu_2$, you should follow the familiar four-step process. We will estimate the difference by comparing the sample means and build an interval around that point estimate by using the standard error of the statistic and a critical *t* value. To construct a two-sample *t* interval for a difference between two means:

- **State** the parameters of interest and the confidence level you will be using to estimate the difference.
- **Plan:** Indicate the type of confidence interval you are constructing and verify that the conditions for Random, Normal, and Independent are satisfied for the two samples.

- **Do** the actual construction of the interval using the following formula:

$$(\bar{x}_1 - \bar{x}_2) \pm t^* \sqrt{\frac{s_1^2}{n_1} + \frac{s_2^2}{n_2}}$$

where $t^*$ is the critical value for the $t$ distribution curve having df = smaller of $n_1 - 1$ and $n_2 - 1$ OR given by technology with area $C$ between $-t^*$ and $t^*$.

- **Conclude** by interpreting the interval in the context of the problem.

**Check for Understanding:** ____ *I can construct and interpret a two-sample* t *interval to compare two means.*

Researchers are interested in determining the effectiveness of a new diet for individuals with heart disease. 200 heart disease patients are selected and randomly assigned to the new diet or the current diet used in the treatment of heart disease. The 100 patients on the new diet lost an average of 9.3 pounds with standard deviation 4.7 pounds. The 100 patients continuing with their current prescribed diet lost an average of 7.4 pounds with standard deviation 4 pounds. Construct and interpret a 95% confidence interval for the difference in mean weight loss for the two diets.

### *Concept 3: Significance tests for $\mu_1 - \mu_2$*

To test a claim about the difference between two population or treatment means, we will follow the same four-step process learned previously. To conduct a two-sample *t* test for the difference between two means:

1) **STATE the parameter of interest and the hypotheses you would like to test.**

   The null hypothesis usually states that the parameters are equal while the alternative hypothesis is that one mean is greater than, less than, or not equal to the other:

   $H_0$: $\mu_1 = \mu_2$

   $H_a$: $\mu_1 > \mu_2$ OR $\mu_1 < \mu_2$ OR $\mu_1 \neq \mu_2$

   State the significance level.

2) **PLAN: Choose the appropriate inference method and check the conditions.**
We must check the Random, Normal, and Independent conditions. Verify that the data come from random samples or the groups in a randomized experiment. To ensure that the sampling distribution of $\bar{x}_1 - \bar{x}_2$ is at least approximately Normal, check that the population distributions are Normal OR $n_1$ and $n_2$ are both at least 30. If sampling is done without replacement, check that the populations are at least 10 times as large as the samples.

3) **DO: If conditions are met, calculate a test statistic and *P*-value.**

$$t = \frac{(\bar{x}_1 - \bar{x}_2) - (\mu_1 - \mu_2)}{\sqrt{\frac{s_1^2}{n_1} + \frac{s_2^2}{n_2}}}$$

and find the *P*-value by calculating the probability of observing a *t* statistic at least this extreme in the direction of the alternative hypothesis. Use the *t* distribution with df = smaller of $n_1 - 1$ and $n_2 - 1$ OR given by technology.

4) **Conclude by interpreting the results of your calculations in the context of the problem.**
If the *P*-value is smaller than the stated significance level, you have significant evidence to reject the null hypothesis. If it is larger than or equal to the significance level, then you fail to reject the null hypothesis.

**Check for Understanding:** ____ *I can perform a two-sample* t *test to compare two means.*

Do boys have better short term memory than girls? A random sample of 200 boys and 150 girls was administered a short term memory test. The average score for boys was 48.9 with standard deviation 12.96. The girls had an average score of 48.4 with standard deviation 11.85. Is there significant evidence at the 5% level to suggest boys have better short term memory than girls? Note: higher test scores indicate better short term memory.

## Chapter Summary: Comparing Two Populations or Groups

In this chapter, you learned the inference procedures that help us compare two parameters. Whether you are comparing two proportions or two means, the processes for constructing a confidence interval or performing a significance test are the same. When comparing two proportions, you use two-sample *z* procedures to reach your conclusions. If you are dealing with means, you use two-sample *t* procedures. Like we learned for one-sample procedures, it is important to identify which procedure you are using, verify that the appropriate conditions are met, perform the necessary calculations, and interpret your results in the context of the problem.

### After You Read: "How Can I Close the Gap?"

Complete the vocabulary puzzle, multiple choice questions, and FRAPPY. Check your answers and your performance on each of the targets.

| Target | Got It! | Almost There | Needs Some Work |
|---|---|---|---|
| I can describe the characteristics of the sampling distribution of $\hat{p}_1 - \hat{p}_2$ | | | |
| I can calculate probabilities using the sampling distribution of $\hat{p}_1 - \hat{p}_2$ | | | |
| I can determine whether or not the conditions for performing inference are met. | | | |
| I can construct and interpret a confidence interval to compare two proportions. | | | |
| I can perform a significance test to compare two proportions. | | | |
| I can interpret the results of inference procedures in a randomized experiment. | | | |
| I can describe the characteristics of the sampling distribution of $\bar{x}_1 - \bar{x}_2$ | | | |
| I can calculate probabilities using the sampling distribution of $\bar{x}_1 - \bar{x}_2$ | | | |
| I can determine whether or not the conditions for performing inference are met. | | | |
| I can construct and interpret a two-sample *t* interval to compare two means. | | | |
| I can perform a two-sample *t* test to compare two means. | | | |
| I can interpret the results of inference procedures in a randomized experiment. | | | |
| I can determine the proper inference procedure to use in a given setting. | | | |

Did you check "Needs Some Work" for any of the targets? If so, what will you do to address your needs for those targets?

*Learning Plan:*

## Chapter 10 Multiple Choice Practice

**Directions.** *Identify the choice that best completes the statement or answers the question. Check your answers and note your performance when you are finished.*

1. Is the proportion of marshmallows in Mr. Miller's favorite breakfast cereal lower than it used to be? To determine this, you test the hypotheses $H_0$: $p_{old} = p_{new}$, $H_a$: $p_{old} > p_{new}$ at the α = 0.05 level. You calculate a test statistic of 1.980. Which of the following is the appropriate P-value and conclusion for your test?

| | |
|---|---|
| A. | *P*-value = 0.047; fail to reject $H_0$; we do not have convincing evidence that the proportion of marshmallows has been reduced. |
| B. | *P*-value = 0.047; accept $H_a$; there is convincing evidence that the proportion of marshmallows has been reduced. |
| C. | *P*-value = 0.024; fail to reject $H_0$; we do not have convincing evidence that the proportion of marshmallows has been reduced. |
| D. | *P*-value = 0.024; reject $H_0$; we have convincing evidence that the proportion of marshmallows has been reduced. |
| E. | *P*-value = 0.024; fail to reject $H_0$; we have convincing evidence that the proportion of marshmallows has not changed. |

2. An SRS of 100 teachers showed that 64 owned smartphones. An SRS of 100 students showed that 80 owned smartphones. Let $p_T$ be the proportion of all teachers who own smartphones, and let $p_S$ be the proportion of all students who own smartphones. A 95% confidence interval for the difference $p_T - p_S$ is

| | |
|---|---|
| A. | (0.264, 0.056) |
| B. | (0.098, 0.222) |
| C. | (-0.222, -0.098) |
| D. | (-0.264, -0.056) |
| E. | (-0.283, -0.038) |

3. A school receives textbooks independently from two suppliers. An SRS of 400 textbooks from supplier 1 finds 20 that are defective. An SRS of 100 textbooks from supplier 2 finds 10 that are defective. Let $p_1$ and $p_2$ be the proportions of all textbooks from suppliers 1 and 2, respectively, that are defective. Which of the following represents a 95% confidence interval for $p_1 - p_2$?

| | |
|---|---|
| A. | $-0.05 \pm 1.96\sqrt{\frac{(0.05)(0.95)}{400} - \frac{(0.1)(0.9)}{100}}$ |
| B. | $-0.05 \pm 1.96\sqrt{\frac{(0.05)(0.95)}{400} + \frac{(0.1)(0.9)}{100}}$ |
| C. | $-0.05 \pm 1.64\sqrt{\frac{(0.05)(0.95)}{400} - \frac{(0.1)(0.9)}{100}}$ |
| D. | $-0.05 \pm 1.64\sqrt{\frac{(0.05)(0.95)}{400} + \frac{(0.1)(0.9)}{100}}$ |
| E. | $-0.05 \pm 1.64\sqrt{\frac{(0.06)(0.94)}{500}}$ |

4. An agricultural researcher wishes to see if a new fertilizer helps increase the yield of tomato plants. One hundred tomato plants in individual containers are randomly assigned to two different groups. Plants in both groups are treated identically, except that the plants in group 2 are sprayed weekly with the fertilizer, while the plants in group 1 are not. After 4 weeks, 12 of the 50 plants in group 1 exhibited an increased yield, and 18 of the 50 plants in group 2 showed an increased yield. Let $p_1$ be the actual proportion of all tomato plants of this variety that would experience an increased yield under the fertilizer treatment, and let $p_2$ be the actual proportion of all tomato plants of this variety that would experience an increased yield under with no fertilizer treatment, assuming that the tomatoes are grown under conditions similar to those in the experiment. Is there evidence of an increase in the proportion of tomato plants with increased yield for those sprayed with fertilizer? To determine this, you test the hypotheses $H_0$: $p_1 = p_2$, $H_a$: $p_1 < p_2$. The *P*-value of your test is

| | |
|---|---|
| A. | greater than 0.10. |
| B. | between 0.05 and 0.10. |
| C. | between 0.01 and 0.05. |
| D. | between 0.001 and 0.01. |
| E. | below 0.001. |

5. An SRS of 45 male employees at a large company found that 36 felt that the company was supportive of female and minority employees. An independent SRS of 40 female employees found that 24 felt that the company was supportive of female and minority employees. Let $p_1$ represent the proportion of all male employees at the company and $p_2$ represent the proportion of all female employees members at the company who hold this opinion. We wish to test the hypotheses $H_0$: $p_1 - p_2 = 0$ *vs.* $H_a$: $p_1 - p_2 < 0$. Which of the following is the correct expression for the test statistic?

| | |
|---|---|
| A. | $\dfrac{0.8-0.6}{\sqrt{\dfrac{(0.8)(0.2)}{45}+\dfrac{(0.6)(0.4)}{40}}}$ |
| B. | $\dfrac{0.8-0.6}{\sqrt{\dfrac{(0.706)(0.294)}{45}+\dfrac{(0.706)(0.294)}{40}}}$ |
| C. | $\dfrac{0.8-0.6}{\sqrt{\dfrac{(0.706)(0.294)}{45}-\dfrac{(0.706)(0.294)}{40}}}$ |
| D. | $\dfrac{0.8-0.6}{\dfrac{(0.8)(0.2)}{\sqrt{45}}+\dfrac{(0.6)(0.4)}{\sqrt{40}}}$ |
| E. | $\dfrac{0.8}{\dfrac{(0.8)(0.2)}{\sqrt{45}}}+\dfrac{0.6}{\dfrac{(0.6)(0.4)}{\sqrt{40}}}$ |

6. Some researchers have conjectured that stem-pitting disease in peach tree seedlings might be controlled with weed and soil treatment. An experiment was conducted to compare peach tree seedling growth with soil and weeds treated with one of two herbicides. In a field containing 20 seedlings, 10 were randomly selected from throughout the field and assigned to receive Herbicide A. The remaining 10 seedlings were to receive Herbicide B. Soil and weeds for each seedling were treated with the

appropriate herbicide, and at the end of the study period, the height (in centimeters) was recorded for each seedling. A box plot of each data set showed no indication of non-Normality. The following results were obtained:

| | $\bar{x}$ (cm) | $s_x$ (cm) |
|---|---|---|
| Herbicide A | 94.5 | 10 |
| Herbicide B | 109.1 | 9 |

Suppose we wished to determine if there is a significant difference in mean height for the seedlings treated with the different herbicides. Based on our data, which of the following is the value of test statistic?

| | |
|---|---|
| A. | 14.60 |
| B. | 7.80 |
| C. | 3.43 |
| D. | 2.54 |
| E. | 1.14 |

7. A researcher wished to test the effect of the addition of extra calcium on the "tastiness" of yogurt. Sixty-two adult volunteers were randomly divided into two groups of 31 subjects each. Group 1 tasted yogurt containing the extra calcium. Group 2 tasted yogurt from the same batch as group 1 but without the added calcium. Both groups rated the flavor on a scale of 1 to 10, with 1 being "very unpleasant" and 10 being "very pleasant." The mean rating for group 1 was 6.5 with a standard deviation of 1.5. The mean rating for group 2 was 7.0 with a standard deviation of 2.0. Let $\mu_1$ and $\mu_2$ represent the true mean ratings we would observe for the entire population represented by the volunteers if all of them tasted, respectively, the yogurt with and without the added calcium. Which of the following would lead us to believe that the *t*-procedures were not safe to use in this situation?

| | |
|---|---|
| A. | The sample medians and means for the two groups were slightly different. |
| B. | The distributions of the data for the two groups were both slightly skewed right. |
| C. | The data are integers between 1 and 10 and so cannot be normal. |
| D. | The standard deviations from both samples were very different from each other. |
| E. | None of the above. |

8. A researcher wishes to compare the effect of two stepping heights (low and high) on heart rate in a step-aerobics workout. The researcher constructs a 98% confidence interval for the difference in mean heart rate between those who did the high and those who did the low stepping heights. Which of the following is a correct interpretation of this interval?

| | |
|---|---|
| A. | 98% of the time, the true difference in the mean heart rate of subjects in the high-step *vs.* low-step groups will be in this interval. |
| B. | We are 98% confident that this interval captures the true difference in mean heart rate of subjects like these who receive the high-step and low-step treatments. |
| C. | There is a 0.98 probability that the true difference in mean heart rate of subjects in the high-step *vs.* low-step groups in this interval. |
| D. | 98% of the intervals constructed in this way will contain the value 0. |
| E. | There is a 98% probability that we have <u>not</u> made a Type I error. |

9. Using the setting from problem 8. The researcher decides to test the hypotheses $H_0$: $\mu_1 - \mu_2 = 0$ *vs.* $H_a$: $\mu_1 - \mu_2 < 0$ at the $\alpha = 0.05$ level and produces a *P*-value of 0.0475. Which of the following is a correct interpretation of this result?

| | |
|---|---|
| A. | The probability that there is a difference is 0.0475. |
| B. | The probability that this test resulted in a Type II error is 0.0475. |
| C. | If this test were repeated many times, we would make a Type I error 4.75% of the time. |
| D. | If the null hypothesis is true, the probability of getting a difference in sample means as far or farther from 0 as the difference in our samples is 0.0475. |
| E. | If the null hypothesis is false, the probability of getting a difference in sample means as far or farther from 0 as the difference in our samples is 0.0475. |

10. The researcher in question 8 randomly assigned 50 adult volunteers to two groups of 25 subjects each. Group 1 did a standard step-aerobics workout at the low height. The mean heart rate at the end of the workout for the subjects in group 1 was 90 beats per minute with a standard deviation of 9 beats per minute. Group 2 did the same workout but at the high step height. The mean heart rate at the end of the workout for the subjects in group 2 was 95.2 beats per minute with a standard deviation of 12.3 beats per minute. Assuming the conditions are met, which of the following could be the 98% confidence interval for the difference in mean heart rates based on these results?

| | |
|---|---|
| A. | (2.15, 8.25) |
| B. | (-0.77, 11.17) |
| C. | (-2.13, 12.54) |
| D. | (-2.16, 12.56) |
| E. | (-4.09, 14.49) |

## Multiple Choice Answers

| Problem | Answer | Concept | Right | Wrong | Simple Mistake? | Need to Study More |
|---|---|---|---|---|---|---|
| 1 | D | Significance Test, *P*-value, and Conclusion | | | | |
| 2 | E | Confidence Interval for Difference in Proportions | | | | |
| 3 | B | Confidence Interval for Difference in Proportions | | | | |
| 4 | B | P-value for Test of Significance | | | | |
| 5 | B | Test Statistic | | | | |
| 6 | C | Significance Test for Difference in Means | | | | |
| 7 | E | Conditions for Inference | | | | |
| 8 | B | Interpret Confidence Interval | | | | |
| 9 | D | Interpret P-value | | | | |
| 10 | D | Confidence Interval for Difference in Means | | | | |

## FRAPPY! Free Response AP Problem, Yay!

The following problem is modeled after actual Advanced Placement Statistics free response questions. Your task is to generate a complete, concise response in 15 minutes. After you generate your response, view two example solutions and determine whether you feel they are "complete," "substantial," "developing" or "minimal." If they are not "complete," what would you suggest to the student who wrote them to increase their score? Finally, you will be provided with a rubric. Score your response and note what, if anything, you would do differently to increase your own score.

Researchers are interested in whether or not women who are part of a prenatal care program give birth to babies with a higher average birth weight than those who do not take part in the program. A random sample of hospital records indicates that the average birth weight for 75 babies born to mothers enrolled in a prenatal care program was 3100 g with standard deviation 420 g. A separate random sample of hospital records indicates that the average birth weight for 75 babies born to women who did not take part in a prenatal care program was 2750 g with standard deviation 425 g. Do these data provide convincing evidence that mothers who participate in a prenatal care program have babies with a higher average birth weight than those who don't?

**Student Response 1:**

We will perform a two-sample t-test for the difference in means.

$H_0$: $\mu_Y = \mu_N$

$H_a$: $\mu_Y > \mu_N$ (Note, Y = enrolled in program, N = not enrolled)

We are told the samples are random. Both sample sizes (75) are greater than 30.

$t = \dfrac{3100 - 2750}{\sqrt{\dfrac{420^2}{75} + \dfrac{425^2}{75}}} = 5.07$ and our *p*-value is 0.000000578.

We have significant evidence at the 5% level to reject the null hypothesis. The babies born to mothers enrolled in the prenatal care program have higher average? birth weights than the babies born to mothers who are not enrolled in the program.

How would you score this response? Is it substantial? Complete? Developing? Minimal? Is there anything this student could do to earn a better score?

**Student Response 2:**

I will construct a 95% confidence interval for the difference in mean weights. We are told we have a random samples and both sample sizes are greater than 30.

$$(3100 - 2750) \pm t^* \sqrt{\frac{420^2}{75} + \frac{425^2}{75}} = (213.65, 486.34)$$

We are 95% confident that the true difference in average? birth weights between babies born to mothers enrolled in the program and babies born to mothers not enrolled in the program is between 213.65 g and 486.34 g. Since 0 is not contained in this interval, we have evidence to suggest babies born to mothers enrolled in the program have an average weight between 213.65 g and 486.34 g greater than the average weight of babies born to mothers not enrolled in the program.

How would you score this response? Is it substantial? Complete? Developing? Minimal? Is there anything this student could do to earn a better score?

## Scoring Rubric

Use the following rubric to score your response. Each part receives a score of "Essentially Correct," "Partially Correct," or "Incorrect." When you have scored your response, reflect on your understanding of the concepts addressed in this problem. If necessary, note what you would do differently on future questions like this to increase your score.

### Intent of the Question

The goal of this question is to determine your ability to conduct a significance test for a single proportion the difference between two means.

### Solution

The solution should contain 4 parts:

**Hypotheses**: Must be stated correctly: $H_0$: $\mu_Y - \mu_N = 0$ and $H_a$: $\mu_Y - \mu_N > 0$ where $\mu_Y$ = true mean birth weight of babies born to mothers enrolled in a prenatal care program and $\mu_N$ = true mean birth weight of babies born to mothers who were not enrolled in a prenatal care program.

**Name of Test and Conditions**: The test must be identified by name or formula as a two-sample *t* test for a difference in population means. Two random samples were obtained. Both sample sizes are greater than 30. Both samples are less than 10% of their respective populations since this is a large hospital.

**Mechanics**: $t = 5.07$ and *p*-value = $5.78 \times 10^{-7}$ with 147.97 degrees of freedom.

**Conclusion**: Because the *p*-value is smaller than any reasonable significance level we reject the null in favor of the alternative hypothesis. There is significant evidence to suggest the mean birth weight of babies born to mothers who participate in a prenatal care program is higher than the weight of those born to mothers who do not participate.

### Scoring

Each element scored as essentially correct (E), partially correct (P), or incorrect (I).

**Hypotheses** are essentially correct if the hypotheses are written correctly. This part is partially correct if the test isn't identified or if the hypotheses are written incorrectly.

**Name and Conditions** is essentially correct if the response correctly identifies the test by name or formula AND correctly checks the Random, Normal, and 10% conditions. This part is partially correct if at least one of the conditions is not checked and incorrect if only one condition is checked.

**Mechanics** is essentially correct if the response correctly calculates the test statistic and *p*-value. This part is partially correct if one of the calculations is incorrect.

**Conclusion** is essentially correct if the response correctly rejects the null hypothesis due to the very small *p*-value and provides an interpretation in context. This part is partially correct if the response fails to justify the decision by indicating that the p-value is very small OR if the conclusion lacks an interpretation in context.

**Scoring:** This problem has four elements, each receiving an E, P, or I. Assign one point to each E, 0.5 points to each P, and 0 points to each I. Total the points to determine your score. If a score falls between two whole values, consider the strength of the entire response to determine whether to round up or down.

# Chapter 10: Comparing Two Populations or Groups

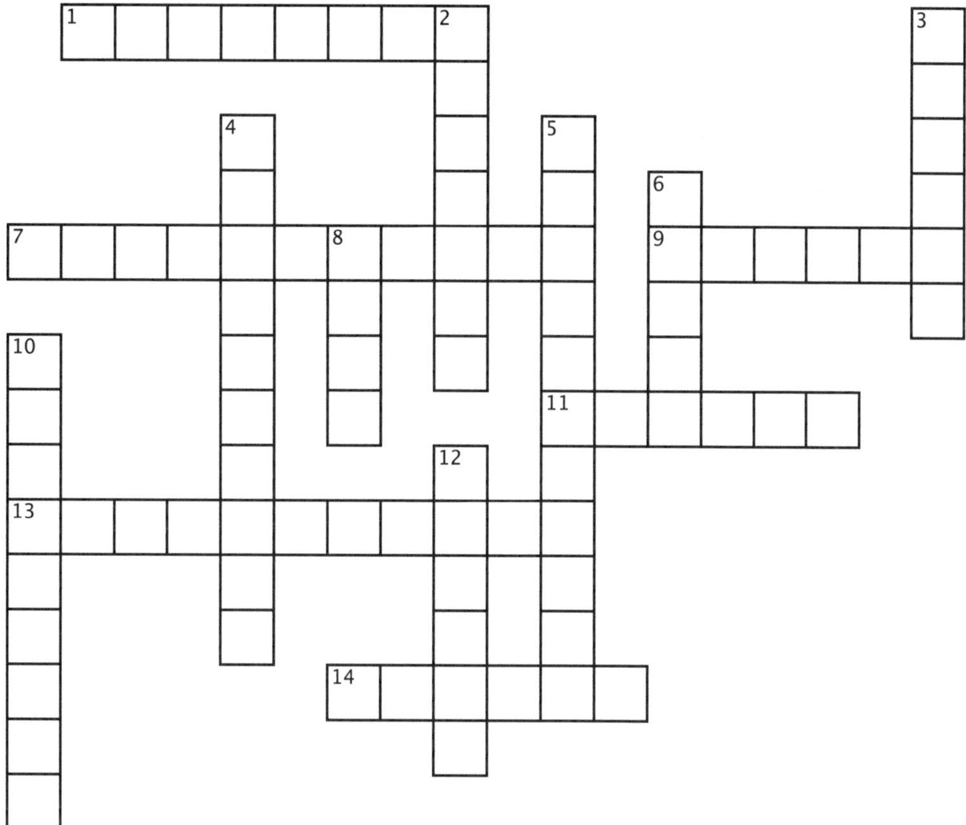

**ross**

an alternative hypothesis the seeks evidence of a difference requires the use of a ____ test (two words)

to perform two-sample procedures, the random samples should be____

procedures that yield accurate results, even when a condition is violated, are ____

if the number of successes and failures from both samples are greater than ten, the sampling distribution for the difference between the two proportions will be approximately ____

results that are too unlikely to be due to chance alone are considered statistically ____

in a two-proportion test, we calculate a ____ proportion

**Down**

2. particular t distributions are distinguished by ____ of freedom
3. the sampling distribution for the difference between means will be approximately Normal if both sample sizes are greater than
4. hypotheses should always be written in terms of the ____
5. the hypthesis we are seeking evidence for
6. the margin of error in a confidence interval consists of a critical value and standard ____ of the statistic
8. the hypothesis we are seeking evidence against
10. to compare two proportions, we can construct a ____ z interval (two words)
12. if we want to compare two population proprtions, we must have data from two ____ samples

# Chapter 11: Inference for Distributions of Categorical Data

*"By a small sample, we may judge the whole piece." Miguel de Cervantes (*Don Quixote*)*

## Chapter Overview

So far, our study of inference has focused on how to estimate and test claims about single means and proportions as well as about differences between means and proportions for quantitative variables. In this chapter, we will shift our focus to inference about distributions of and relationships between categorical variables. You will learn how to perform three different significance tests for distributions of categorical data, allowing you to determine (1) whether or not a sample distribution differs significantly from a hypothesized distribution, (2) whether or not the distribution of a categorical variable differs between multiple populations, or (3) whether or not an association exists between two categorical variables. As before, each test will consist of hypotheses, conditions to be checked, a test statistic and *P*-value, and a conclusion in context. It is easy to confuse the three tests, so be sure to study the differences between them so you know when to use each one!

## Sections in this Chapter



## Plan Your Learning

Use the following *suggested* guide to help plan your reading and assignments. Note: your teacher may schedule a different pacing. Be sure to follow his or her instructions!

| **Read** | 11.1: pp 676-684 | 11.1: pp 685-691 | 11.2: pp 696-712 |
|---|---|---|---|
| **Do** | 1, 3, 5 | 7, 9, 11, 17 | 19-22, 27, 29, 31, 33, 35, 43 |

| **Read** | 11.2: pp 713-724 | Chapter Summary |
|---|---|---|
| **Do** | 45, 49, 51, 53-58 | Multiple Choice FRAPPY! |

## Section 11.1: Chi-Square Goodness-of-Fit Tests

### Before You Read: Section Summary

In this section, you will be introduced to the chi-square distributions and the chi-square test statistic. The chi-square statistic provides us with a way to measure the differences between an observed and a hypothesized distribution of categorical data. When certain conditions are satisfied, we can model the sampling distribution of this statistic with a chi-square distribution and calculate *P*-values. We can use the chi-square goodness-of-fit test to determine whether or not a significant difference between an observed and hypothesized distribution of a categorical variable exists. Finally, we can use a follow-up analysis to determine which categories contributed the most to the difference.

### "Where Am I Going?"

### Learning Targets:

- ____ I can compute expected counts and contributions to the chi-square statistic.
- ____ I can check the Random, Large Sample Size, and Independent conditions before performing a chi-square test.
- ____ I can use a chi-square goodness-of-fit test to determine whether sample data are inconsistent with a specified distribution of a categorical variable.
- ____ I can examine individual components of the chi-square statistic as part of a follow-up analysis.

### While You Read: Key Vocabulary and Concepts

one-way table:

chi-square goodness-of-fit test:

observed counts:

expected counts:

chi-square statistic:

chi-square distribution:

large sample size condition:

components:

**After You Read: "Where Am I Now?"**
**Check for Understanding**

### *Concept 1: The Chi-Square Test Statistic and Distributions*

When testing a claim about a distribution of categorical data, we are interested in knowing how the observed counts from the sample compare to the counts that would be expected if the null hypothesis were true. Like we did with means and proportions, we will standardize the difference between observed and expected values using a test statistic. To determine the chi-square test statistic for a table of observed counts, use the following formula:

$$x^2 = \sum \frac{(observed - expected)^2}{expected}$$

This test statistic can then be used with the appropriate chi-square distribution to determine a *P*-value. If the observed values are far from the expected values, our test statistic will be large, giving evidence against the null hypothesis. The chi-square distributions are a family of right-skewed distributions that are defined by degrees of freedom based on the number of categories the variable takes on. We can use a table or technology to find a *P*-value for a particular $\chi^2$ value.

**Check for Understanding:** ____*I can compute expected counts and contributions to the chi-square statistic.*

After playing a dice game with a friend, you suspect the die may not be fair. That is, you suspect some numbers may be rolled more often than you would expect. To test your suspicion, you roll the die 300 times and record the results.

| Value: | 1 | 2 | 3 | 4 | 5 | 6 |
|---|---|---|---|---|---|---|
| Frequency: | 42 | 55 | 38 | 57 | 64 | 44 |

Use these observed counts to determine the chi-square statistic for this example.

### *Concept 2: Chi-Square Goodness-of-Fit Test*

To perform a chi-square goodness-of-fit test for a claim about a population distribution of categorical data, we will follow the same basic process as we did for means and proportions. That is, to test a claim about the population distribution of a categorical variable:

1) **STATE the hypotheses you would like to test.**
   When testing a claim about a distribution of a categorical variable, we start by defining hypotheses. The null hypothesis assumes that the hypothesized distribution

is correct as stated while the alternative hypothesis is that the specified distribution is different:

$H_0$: *The specified distribution of the categorical variable is correct.*

$H_a$: *The specified distribution of the categorical variable is not correct.*

State a significance level.

2) **PLAN: Choose the appropriate inference method and check the conditions.**
We must check the conditions to ensure that we can use a chi-square distribution to determine our *P*-value. The data must come from a random sample or randomized experiment. All expected counts must be at least 5. Finally, individual observations must be independent—when sampling without replacement, check that the population is at least 10 times as large as the sample.

3) **DO: If conditions are met, calculate a test statistic and *P*-value.**
Compute the chi-square test statistic $x^2 = \sum \frac{(observed - expected)^2}{expected}$ and find the *P*-value by using a chi-square distribution with *k* - 1 degrees of freedom (where *k* is the number of categories).

4) **CONCLUDE by interpreting the results of your calculations in the context of the problem.**
If the *P*-value is smaller than the stated significance level, you can conclude that you have significant evidence to reject the null hypothesis. If the *P*-value is larger than or equal to the significance level, then you fail to reject the null hypothesis.

If you find evidence to reject the null hypothesis, you should perform a follow up analysis to determine the components that contributed the most to the chi-square statistic. That is, determine which observed categories differed the most from the expected values.

**Check for Understanding:** ____ *I can check the Random, Large Sample Size, and Independent conditions before performing a chi-square test,* ____ *I can use a chi-square goodness-of-fit test to determine whether sample data are inconsistent with a specified distribution of a categorical variable, and* ____ *I can examine individual components of the chi-square statistic as part of a follow-up analysis.*

Use the observed counts from the previous Check for Understanding to determine whether or not you have convincing evidence that the die is not fair. If you have significant evidence, perform a follow-up analysis.

## Section 11.2: Inference for Relationships

### Before You Read: Section Summary

Chi-square goodness-of-fit tests allow us to compare the distribution of one categorical variable to a hypothesized distribution. However, sometimes we are interested in comparing the distribution of a categorical variable across several populations or treatments. Chi-square tests for homogeneity allow us to do this. In this section you will learn how to conduct the chi-square test for homogeneity and how to perform a follow-up analysis, just like you did for the goodness-of-fit test. Finally, you will learn how to conduct a chi-square test for association/independence to determine whether or not there is convincing evidence that two categorical variables are related. While this test is the same as the test for homogeneity in its mechanics, the hypotheses are different. Be sure to note the distinction!

### "Where Am I Going?"

**Learning Targets:**

- ____ I can use a chi-square test for homogeneity to determine whether or not the distribution of a categorical variable differs for several populations or treatments.
- ____ I can use a chi-square test of association/independence to determine whether there is convincing evidence of an association between two categorical variables.
- ____ I can distinguish between the three types of chi-square tests.

### While You Read: Key Vocabulary and Concepts

multiple comparisons:

expected counts:

chi-square test for homogeneity:

chi-square test of association/independence:

### After You Read: "Where Am I Now?" Check for Understanding

#### *Concept 1: Expected Counts and the Chi-Square Statistic*

While a one-way table can summarize data for a single categorical variable, a two-way table can be used to summarize data on the relationship between two categorical variables. When these data are produced using independent random samples from several populations or from a randomized comparative experiment, we can test whether or not the actual distribution of the categorical variable is the same for each population or treatment. If the data are produced using a single random sample from a population and classified according to two categorical variables, we can test whether or not there is a relationship between those variables. In each case, we must compare the observed counts to expected counts. To determine an expected count, we use the general formula $expected = \frac{row\ total \bullet column\ total}{table\ total}$. Then, just like we did with the goodness-of-fit test, we calculate the chi-square statistic using $x^2 = \sum \frac{(observed - expected)^2}{expected}$.

### *Concept 2: Chi-Square Test for Homogeneity*

To determine whether or not a distribution of a categorical variable differs for two or more populations or treatments, we use a chi-square test for homogeneity. This test is similar to the goodness-of-fit test in the sense that we compare observed counts to expected counts using a chi-square test statistic. However, the hypotheses and degrees of freedom differ slightly.

1) **STATE the hypotheses you would like to test.**
   When testing a claim about the distribution of a single categorical variable for two or more populations or treatments, we start by defining hypotheses. The null hypothesis says that the variable has the same distribution for all of the populations or treatments. The alternative hypothesis is that the distribution of that variable is not the same for all of the populations or treatments:
   - $H_0$: There is no difference in the distribution of a categorical variable for several populations or treatments.
   - $H_a$: There is a difference in the distribution of a categorical variable for several populations or treatments.

2) **PLAN: Choose the appropriate inference method and check the conditions.**
   We must check the conditions to ensure that we can use a chi-square distribution to determine our *P*-value. The data must come from a random sample or randomized experiment. All expected counts must be at least 5. Finally, individual observations must be independent—when sampling without replacement, check that the population is at least 10 times as large as the sample.

3) **DO: If conditions are met, calculate a test statistic and *P*-value.**
   Compute the chi-square test statistic $x^2 = \sum \frac{(observed - expected)^2}{expected}$ and find the *P*-value by using a chi-square distribution with *(number of rows - 1)(number of columns – 1)* degrees of freedom.

4) **CONCLUDE by interpreting the results of your calculations in the context of the problem.**
   If the *P*-value is smaller than the stated significance level, you can conclude that you have significant evidence to reject the null hypothesis. If *the* P-value is larger than or equal to the significance level, then you fail to reject the null hypothesis.

If you find sufficient evidence to reject the null hypothesis, you should perform a follow up analysis to determine the components that contributed the most to the chi-square statistic. That is, determine which observed categories differed the most from the expected values.

**Check for Understanding:** ____ *I can use a chi-square test for homogeneity to determine whether the distribution of a categorical variable differs for several populations or treatments.*

A recent study tracked the television viewing habits of 100 randomly selected first-grade boys and 200 randomly selected first-grade girls. Each child was asked to identify their favorite TV show. The following table summarizes the results:

| | Zooboomafoo | iCarly | Phineas and Ferb |
|---|---|---|---|
| Boys | 20 | 30 | 50 |
| Girls | 70 | 80 | 50 |

Do these data provide convincing evidence that television preferences differ significantly for boys and girls?

## *Concept 3: Chi-Square Test of Association/Independence*

To determine whether or not two categorical variables are related in a population, we can use a chi-square test of association/independence. The mechanics of this test are exactly the same as those for the test for homogeneity. The only difference is that the hypotheses are defined in terms of an association between the two categorical variables.

$H_0$: There is no association between two categorical variables in the population of interest.

$H_a$: There is an association between two categorical variables in the population of interest.

**Check for Understanding:** ____ *I can use a chi-square test of association/independence to determine whether there is convincing evidence of an association between two categorical variables.*

A recent study looked into the relationship between political views and opinions about nuclear energy. A survey administered to 100 randomly selected adults asked their political leanings as well as their approval of nuclear energy. The results are below:

| | Liberal | Conservative | Independent |
|---|---|---|---|
| Approve | 10 | 15 | 20 |
| Disapprove | 9 | 2 | 16 |
| No Opinion | 8 | 2 | 18 |

Do these data provide convincing evidence that political leanings and views on nuclear energy are associated in the larger population of adults from which the sample was selected?

## Chapter Summary: Inference for Distributions of Categorical Data

In this chapter, you learned the inference procedures for distributions of categorical data. A chi-square goodness-of-fit test can be used to determine whether or not an observed distribution of a categorical variable differs from a hypothesized distribution. When examining the distribution of a single categorical variable in multiple populations or treatments, a test for homogeneity can be used to determine whether the distributions differ. Finally, we can use a test for association/independence to determine whether or not two categorical variables are related in a population. In each test, we use the chi-square test statistic to measure how much the observed counts differ from the expected counts. When our test provides significant evidence against the null hypothesis, we can use a follow-up analysis to determine which component(s) contributed the most to the test statistic.

### After You Read: "How Can I Close the Gap?"

Complete the vocabulary puzzle, multiple choice questions, and FRAPPY. Check your answers and your performance on each of the targets.

| Target | Got It! | Almost There | Needs Some Work |
|---|---|---|---|
| I can compute expected counts, conditional distributions, and contributions to the chi-square statistic. | | | |
| I can check the Random, Large Sample Size, and Independent conditions before performing a chi-square test. | | | |
| I can use a chi-square goodness-of-fit test to determine whether sample data are inconsistent with a specified distribution of a categorical variable. | | | |
| I can examine individual components of the chi-square statistic as part of a follow-up analysis. | | | |
| I can use a chi-square test for homogeneity to determine whether the distribution of a categorical variable differs for several populations or treatments. | | | |
| I can use a chi-square test of association/independence to determine whether there is convincing evidence of an association between two categorical variables. | | | |
| I can distinguish between the three types of chi-square tests. | | | |

Did you check "Needs Some Work" for any of the targets? If so, what will you do to address your needs for those targets?

*Learning Plan:*

## Chapter 11 Multiple Choice Practice

**Directions.** *Identify the choice that best completes the statement or answers the question. Check your answers and note your performance when you are finished.*

1. To test the effectiveness of your calculator's random number generator, you randomly select 1000 numbers from a standard Normal distribution. You classify these 1000 numbers according to whether their values are at most –2, between –2 and 0, between 0 and 2, or at least 2. The results are given in the following table. The expected counts, based on the 68-95-99.7 rule, are given as well.

| | At most -2 | Between -2 and 0 | Between 0 and 2 | At least 2 |
|---|---|---|---|---|
| Observed Count | 18 | 492 | 468 | 22 |
| Expected Count | 25 | 475 | 475 | 25 |

To test to see if the distribution of observed counts differs significantly from the distribution of expected counts, we can use a $\chi^2$ goodness of fit test. For this test, the test statistic has approximately a $\chi^2$ distribution. How many degrees of freedom does this distribution have?

| | |
|---|---|
| A. | 3 |
| B. | 4 |
| C. | 7 |
| D. | 999 |
| E. | 1000 |

2. Which of the following is the component of the $\chi^2$ statistic corresponding to the category "at most −2"?

| | |
|---|---|
| A. | (43)(1000)/2000 |
| B. | (43)(25)/1000 |
| C. | 18/1000 |
| D. | $(18-25)^2/25$ |
| E. | $(18-25)^2/18$ |

3. Which of the following statements is true of chi-square distributions?

| | |
|---|---|
| A. | As the number of degrees of freedom increases, their density curves look less and less like a normal curve. |
| B. | As the number of degrees of freedom increases, their density curves look more and more like a uniform distribution. |
| C. | Their density curves are skewed to the left. |
| D. | They take on only positive values. |
| E. | All of the above are true. |

4. A student at a large high school suspects that Mr. Andreasen is grading his students too harshly. Over the past 10 years the proportions of students in *all* sections of statistics (taught by many different teachers) received grades of A, B, C, D, or F in the following proportions: A: 0.20; B: 0.30; C: 0.30; D: 0.10; and F: 0.10. An SRS of 90 students who took statistics with Mr. Andreasen in the past 10 years produces the following information:

| Grade | A | B | C | D | F |
|---|---|---|---|---|---|
| Number of students | 12 | 26 | 28 | 15 | 9 |

Which of the following conditions must be met before the student can use the $\chi^2$ procedure in this situation?

| | |
|---|---|
| A. | The distribution of grades in all introductory statistics courses must be approximately Normal. |
| B. | The number of categories is small relative to the number of observations. |
| C. | All the observed counts are greater than 5. |
| D. | Each observation was randomly selected from the population of all students. |
| E. | All expected counts are approximately equal. |

5. Which of the following expressions represents the expected count of the grade category D?

| A. | 90/5 |
|---|---|
| B. | (0.10)(90) |
| C. | (0.1)(15) |
| D. | $15^2/90$ |
| E. | $(15-9)^2/9$ |

6. Anne wants to know if males and females prefer different brands of frozen pizzas. She bakes four dozen pizzas made by each of four manufacturers, which she labels brands A, B, C, and D. She then selects a simple random sample of 48 students, records their gender, gives them one slice of each brand and asks which brand they like best. Here are her results:

| | A | B | C | D | Total |
|---|---|---|---|---|---|
| Males | 2 | 4 | 6 | 7 | 19 |
| Females | 11 | 5 | 6 | 7 | 29 |
| Total | 13 | 9 | 12 | 14 | 48 |

If we want to compare the conditional distributions for preferred pizza brand among males to the same distribution for females, which of the following is an appropriate graph to use?

| A. | Parallel dotplots |
|---|---|
| B. | Back-to-back stemplots |
| C. | Segmented bar graphs |
| D. | Side-by-side bar graphs |
| E. | Scatterplot |

7. The appropriate null hypothesis for Anne's question in this problem is:

| A. | There is an association between gender and preferred frozen pizza. |
|---|---|
| B. | Gender and pizza preference are independent. |
| C. | The distribution of preferred pizza for each gender is different. |
| D. | The observed count in each cell is equal to the expected count. |
| E. | The males and females subjects in this experiment have the same distribution of pizza brand preference. |

8. Are the conditions for a chi-square test of association/independence met?

| A. | Yes, because the sample size is greater than 30. |
|---|---|
| B. | Yes, because a simple random sample was selected. |
| C. | No, because the distribution for each gender is different. |
| D. | No, because not all observed counts are greater than 5. |
| E. | No, because 25% of expected counts are less than 5. |

9. Below is a table of individual components of the chi-square test of association/independence for a study done on amount of time spent at a computer and whether or not a person wears glasses:

| | | Wear Glasses? | |
|---|---|---|---|
| | | Yes | No |
| Amount of Computer Screen Time | Above Average | 8.7 | 6.3 |
| | Average | 0.5 | 0.3 |
| | Below Average | 3.1 | 2.2 |

Which of the following statements is supported by the information in this table?

| A. | Above-average screen time individuals wore glasses much less often than expected. |
|---|---|
| B. | Average screen time individuals wore glasses much less often than expected. |
| C. | Below-average screen time individuals wore glasses about as often as expected. |
| D. | You can't determine this without the original observed counts. |
| E. | The chi-square statistic for this test is about 3.5. |

10. A random sample of 200 Canadian students were asked about their hand dominance and whether

they suffer from allergies. Here are the results:

| | | Allergies? | |
|---|---|---|---|
| | | Yes | No |
| Hand dominance | Ambidextrous | 12 | 7 |
| | Left-handed | 11 | 9 |
| | Right-handed | 95 | 66 |

What can you conclude about the relationship between hand dominance and allergies?

| | |
|---|---|
| A. | Using a test for association/independence, there is not enough evidence ($P = 0.13$) to conclu that there is a relationship between hand dominance and allergies. |
| B. | Using a test for association/independence, there is enough evidence ($P = 0.87$) to conclude t there is a relationship between hand dominance and allergies. |
| C. | Using a test for association/independence, there is not enough evidence ($P = 0.87$) to conclu that there is a relationship between hand dominance and allergies. |
| D. | Using a test for association/independence, there is not enough evidence ($P = 0.13$) to conclu that there is a relationship between hand dominance and allergies. |
| E. | We cannot perform a chi-square test on these data. |

## Multiple Choice Answers

| Problem | Answer | Concept | Right | Wrong | Simple Mistake? | Need to Study More |
|---|---|---|---|---|---|---|
| 1 | A | Degrees of Freedom | | | | |
| 2 | D | Chi-square Components | | | | |
| 3 | D | Chi-square Distribution Characteristics | | | | |
| 4 | D | Conditions for Chi-square Procedures | | | | |
| 5 | B | Expected Counts | | | | |
| 6 | D | Conditional Distributions | | | | |
| 7 | B | Chi-square test of Independence | | | | |
| 8 | E | Conditions for Chi-Square Procedures | | | | |
| 9 | D | Follow-up Analysis | | | | |
| 10 | C | Chi-square Conclusions | | | | |

## FRAPPY! Free Response AP Problem, Yay!

The following problem is modeled after actual Advanced Placement Statistics free response questions. Your task is to generate a complete, concise response in 15 minutes. After you generate your response, view two example solutions and determine whether you feel they are "complete," "substantial," "developing" or "minimal." If they are not "complete," what would you suggest to the student who wrote them to increase their score? Finally, you will be provided with a rubric. Score your response and note what, if anything, you would do differently to increase your own score.

A study was performed to determine whether or not the name of a course had an effect on student registrations. A statistics course in a large school district was given 4 different names in a course catalog. Each name corresponded to the exact same statistics course. A random sample of student registrations was recorded and the results are given below:

| Course Name | Number of Registrations |
|---|---|
| Statistical Applications | 25 |
| Statistical Reasoning | 22 |
| Statistical Analysis | 30 |
| The Practice of Statistics | 40 |
| Total | 117 |

Do these data suggest the name of the course has an effect on student registrations? Conduct an appropriate statistical test to support your conclusion.

**Student Response 1:**

It is obvious there is a difference. If there wasn't, all of the course names would have received about 30 registrations. However, "The Practice of Statistics" was by far the most popular course name, earning 10 more registrations than expected. "Statistical Reasoning" is the least popular, earning 8 fewer registrations than expected. It appears course name does have an effect on student registrations.

How would you score this response? Is it substantial? Complete? Developing? Minimal? Is there anything this student could do to earn a better score?

**Student Response 2:**

We will perform a chi-square goodness-of-fit test.

$H_o$: the distribution of registrations is uniform.
$H_a$: the distribution of registrations is not uniform.

Conditions. We are told the data comes from a random sample. All observed registration counts are greater than 5. We will assume the sample is less than 10% of all registrations.

$$x^2 = \frac{(25-29.25)^2}{29.25} + \frac{(22-29.25)^2}{29.25} + \frac{(30-29.25)^2}{29.95} + \frac{(40-29.95)^2}{29.95} = 6.38$$

df = 4 – 1 =3 *P*-value = 0.09

Since the *P*-value is greater than the 5% rule of thumb, we fail to reject the null hypothesis. We do not have evidence that the name of the course has an effect on student registrations.

How would you score this response? Is it substantial? Complete? Developing? Minimal? Is there anything this student could do to earn a better score?

## Scoring Rubric

Use the following rubric to score your response. Each part receives a score of "Essentially Correct," "Partially Correct," or "Incorrect." When you have scored your response, reflect on your understanding of the concepts addressed in this problem. If necessary, note what you would do differently on future questions like this to increase your score.

### Intent of the Question

The primary goals of this question are to assess your ability to (1) state the appropriate hypotheses; (2) identify and compute the appropriate test statistic; (3) make a conclusion in the context of the problem;

### Solution

The solution should contain 4 parts:

Test and Hypotheses: The test must be identified by name or formula as a chi-square goodness-of-fit test and hypotheses must be stated appropriately. $H_0$: *student registrations do not differ by course name* and $H_a$: *student registrations do differ by course name*.

Conditions: Random sample is given. All expected counts are 29.25, which is at least 5. Independence: we can assume that the sample is less than 10% of all registrations for the course.

Mechanics: $\chi^2$ = 6.38 and *P*-value = 0.094 with 3 degrees of freedom. Conclusion: Because the *P*-value is larger than a significance level of 5%, we do not have significant evidence to suggest the course name has an effect on the number of registrations.

### Scoring:

Each element scored as essentially correct (E), partially correct (P), or incorrect (I).

**Name and Hypotheses** is essentially correct if the response correctly identifies the test by name or formula and writes hypotheses correctly. This part is partially correct if the test isn't identified or if the hypotheses are written incorrectly.

**Conditions** is essentially correct if the response correctly checks the Random, Large Sample Size, and 10% conditions. This part is partially correct if one of the conditions is not checked and incorrect if only one condition is checked.

**Mechanics** is essentially correct if the response correctly calculates the test statistic and *P*-value. This part is partially correct if one of the calculations is incorrect.

**Conclusion** is essentially correct if the response correctly fails to reject the null hypothesis because the *P*-value is greater than a significance level of 5% and provides an interpretation in context. This part is partially correct if the response fails to justify the decision by comparing the *P*-value to a significance level OR if the conclusion lacks an interpretation in context.

**Scoring**: This problem has four elements, each receiving an E, P, or I. Assign one point to each E, 0.5 points to each P, and 0 points to each I. Total the points to determine your score. If a score falls between two whole values, consider the strength of the entire response to determine whether to round up or down.

# Chapter 11: Inference for Distributions of Categorical Data

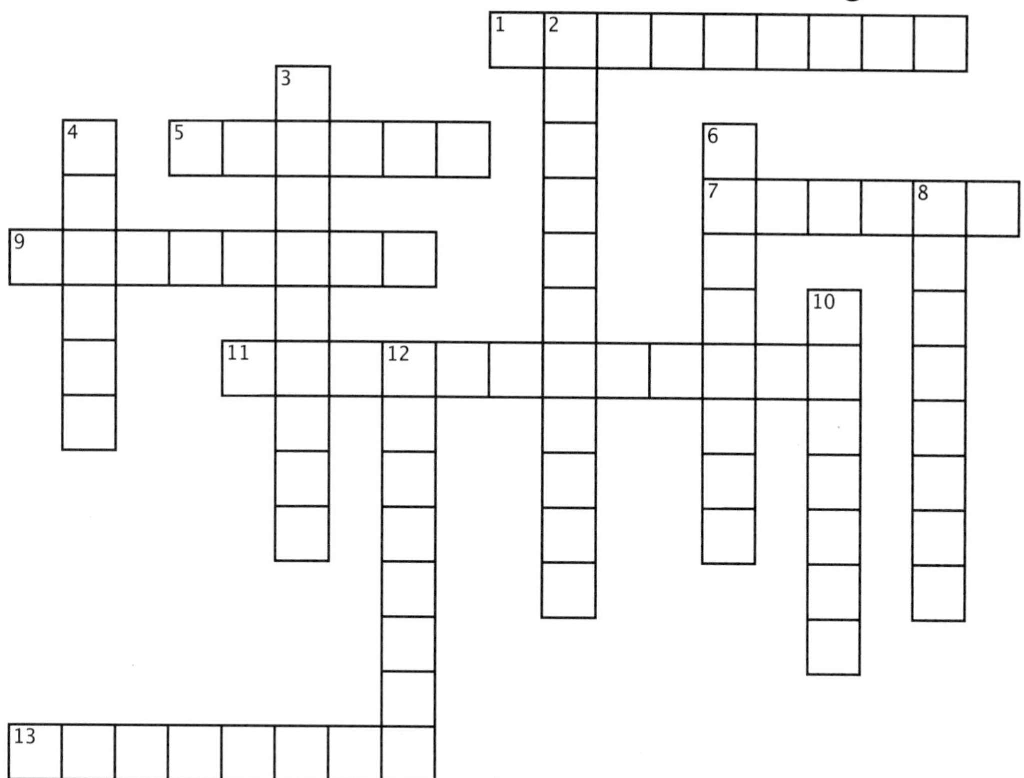


### oss

test statistic used to test hypotheses about distributions of categorical data (two words)

when comparing two or more categorical variables, or one categorical variable over multiple groups, we arrange our data in a _____ table (two words)

a _____ table summarizes the distribution of a single categorical variable (two words)

the problem of doing many comparisons at once with an overall measure of confidence is the problem of _____ comparisons

to test whether there is an association between two categorical variables, use a chi-square test of association / _____

type of categorical count gathered from a sample of data

### Down

2. to test whether there is a difference in the distribution of a categorical variable over several populations or treatments, use a chi-square test of _____
3. a follow up analysis involves identifying the _____ that contribued the most to the chi-square statistic
4. when calculating chi-square statistics, observations should be expressed in _____, not percents
6. _____ of fit test: used to determine whether a population has a certain hypothesized distribution of proportions for a categorical variable
8. after conducting a chi-square test, be sure to carry out a follow-up _____
10. specific chi-square distributions are distinguished by _____ of freedom
12. type of categorical count that we would see if the null hypothesis were true

# Chapter 12: More About Regression

*"A judicious man looks on statistics not to get knowledge, but to save himself from having ignorance foisted on him." Thomas Carlyle*

## Chapter Overview

In the last chapter, you learned how to determine whether or not there was convincing evidence of a relationship between two categorical variables. In this chapter, you will learn how to determine whether or not there is convincing evidence of a relationship between two quantitative variables. You will first learn how to perform inference about the slope of a least-squares regression line. By the end of the first section, you will know how to construct and interpret a confidence interval for the slope as well as how to perform a significance test about it. In the second section, you will learn how to transform data to achieve linearity when a scatterplot shows a curved relationship between two quantitative variables. It has been a while since you studied least-squares regression. This chapter will refresh your memory on some of the key concepts of regression while introducing you to some new inference techniques. As you wrap up your studies and begin your preparations for the AP exam, it might be helpful to look back at the learning targets throughout this guide to determine what topics are in need of a little extra review!

## Sections in this Chapter

**Section 12.1**: Inference for Linear Regression
**Section 12.2**: Transforming to Achieve Linearity

## Plan Your Learning

Use the following *suggested* guide to help plan your reading and assignments. Note: your teacher may schedule a different pacing. Be sure to follow his or her instructions!

| **Read** | 12.1: pp 738-744 | 12.1: pp 744-750 | 12.1: pp 751-758 | 12.2: pp 765-771 |
|---|---|---|---|---|
| **Do** | 1, 3 | 5, 7, 9, 11 | 13, 15, 17, 19 | 21-26, 33, 35 |

| **Read** | 12.2: pp 771-785 | Chapter Summary |
|---|---|---|
| **Do** | 37, 39, 41, 45-48 | Multiple Choice<br>FRAPPY! |

## Section 12.1: Inference for Regression

### Before You Read: Section Summary

When you construct a least-squares regression line based on sample data, your line is an approximation of the population (true) regression line. If you took a different sample and constructed a least-squares regression line, it is very likely you would end up with a slightly different line. In this section, you will learn about the sampling distribution of the sample slope *b* so that you can perform inference about the true slope of the regression line. Like you did with other inference procedures, you will start by learning how to check the conditions for performing inference. Then you will learn how to construct and interpret a confidence interval for the true slope, and how to perform a significance test about the true slope. You will also be re-introduced to computer output for a linear regression analysis. It is important that you be able to interpret this output as a number of AP exam questions involve standard computer output for regression.

### "Where Am I Going?"

**Learning Targets:**

- ____ I can check conditions for performing inference about the slope of the population (true) regression line.
- ____ I can interpret computer output from a least-squares regression analysis.
- ____ I can construct and interpret a confidence interval for the slope of a least-squares regression line.
- ____ I can perform a significance test about the slope of a population regression line.

### While You Read: Key Vocabulary and Concepts

sample regression line:

population (true) regression line:

sampling distribution of *b*:

conditions for regression inference:

standard error of the slope:

*t* interval for the slope:

*t* test for the slope:

**After You Read: "Where Am I Now?"**
**Check for Understanding**

### *Concept 1: The Sampling Distribution of b*

If the conditions for performing inference are met, we can use the slope *b* of the sample regression line $\hat{y} = a + bx$ to estimate or test a claim about the slope $\beta$ of the population (true) regression line $\mu_y = \alpha + \beta x$. The required conditions are:

- **Linear**: There is a true linear relationship between the variables given by $\mu_y = \alpha + \beta x$.
*Is a scatterplot of the sample data linear?*

- **Independent**: Individual observations are independent.
*If sampling without replacement, is the sample size less than 10% of the population?*

- **Normal**: At each *x*-value, the responses *y* follow a Normal distribution.
*Does a graph of the residuals exhibit strong skewness or other signs of non-Normality?*

- **Equal variance**: At each *x*-value, the standard deviation $\sigma$ of the responses *y* is the same.
*Is the amount of scatter above and below 0 on the residual plot roughly the same from the smallest to largest x-values?*

- **Random**: The data are produced by a random sample or randomized experiment.

If we take repeated samples of the same size *n* from a population with a true regression line $\mu_y = \alpha + \beta x$ and determine the sample regression line $\hat{y} = a + bx$, the sampling distribution of the slope *b* will have the following characteristics:

1) The shape of the sampling distribution will be approximately Normal.
2) The mean of the sampling distribution is $\mu_b = \beta$.
3) The standard error of the sampling distribution is $SE_b = \dfrac{s}{s_x\sqrt{n-1}}$ where *s* is the standard deviation of the residuals.

To perform inference about the slope, we note that the sampling distribution of the standardized slope values has a *t* distribution with $n - 2$ degrees of freedom.

### *Concept 2: Confidence Intervals for $\beta$*

When we construct a least-squares regression line $\hat{y} = a + bx$ and the conditions noted above are met, *b* is our estimate of the true slope $\beta$. A level *C* confidence interval for the true slope has the form $b \pm t^*SE_b$ where $t^*$ is the critical value for the *t* distribution with $n - 2$ degrees of freedom that has area *C* between $-t^*$ and $t^*$.

**Check for Understanding:** ____ *I can construct and interpret a confidence interval for the slope of a least-squares regression line.*

A study by *Consumer Reports* rated 10 randomly selected cereals on a 100 point scale (higher numbers are better) and recorded the number of grams of sugar in each serving. The data from the study are below:

| Sugar | 6.0 | 8.0 | 5.0 | 0.0 | 8.0 | 10.0 | 14.0 | 8.0 | 6.0 | 5.0 |
|---|---|---|---|---|---|---|---|---|---|---|
| Rating | 68.40 | 33.98 | 59.43 | 93.70 | 34.38 | 29.51 | 33.17 | 37.04 | 49.12 | 53.31 |

The LSRL for this data is $\hat{r} = 82.62 - 4.77(s)$ where $r$ = rating and $s$ = sugar.
Construct and interpret a 95% confidence interval for the true slope of the regression line. Assume the conditions for performing inference are met.

### *Concept 2: Significance Test for β*

When the conditions for inference are met, not only can we estimate the true slope from *b*, we can also test whether or not a specified value for $\beta$ is plausible. The process for testing a claim about a population (true) slope follows the same format used for other significance tests.

1) **STATE the parameter of interest and the hypotheses you would like to test.**
   The null hypothesis states that the population (true) slope is equal to a particular value while the alternative hypothesis is that the population (true) slope is greater than, less than, or not equal to that value:
   $$H_0: \beta = \beta_0$$
   $$H_a: \beta > \beta_0 \text{ OR } \beta < \beta_0 \text{ OR } \beta \neq \beta_0$$
   State a significance level.
2) **PLAN: Choose the appropriate inference method and check the conditions.**
   Check the linear, independent, Normal, equal variance, and random conditions.
3) **DO: If conditions are met, calculate a test statistic and *P*-value.**
   Compute the *t* test statistic
   $$t = \frac{b - \beta_0}{SE_b}$$
   and find the *P*-value by calculating the probability of observing a *t* statistic at least this extreme in the direction of the alternative hypothesis in a *t* distribution with $n - 2$ degrees of freedom.
4) **CONCLUDE by interpreting the results of your calculations in the context of the problem.**

If the *P*-value is smaller than the stated significance level, you can conclude that you have sufficient evidence to reject the null hypothesis. If the *P*-value is larger than the significance level, then you fail to reject the null hypothesis.

Generally, we test whether or not the true slope is zero, which would indicate no linear relationship between *x* and *y*. Like other inference situations, a significance test can tell us whether or not a claim about the parameter is plausible, while using a confidence interval can give us additional information about its true value.

**Check for Understanding:** ____ *I can perform a significance test about the slope of a population (true) regression line and* ____ *I can interpret computer output from a least-squares regression analysis.*

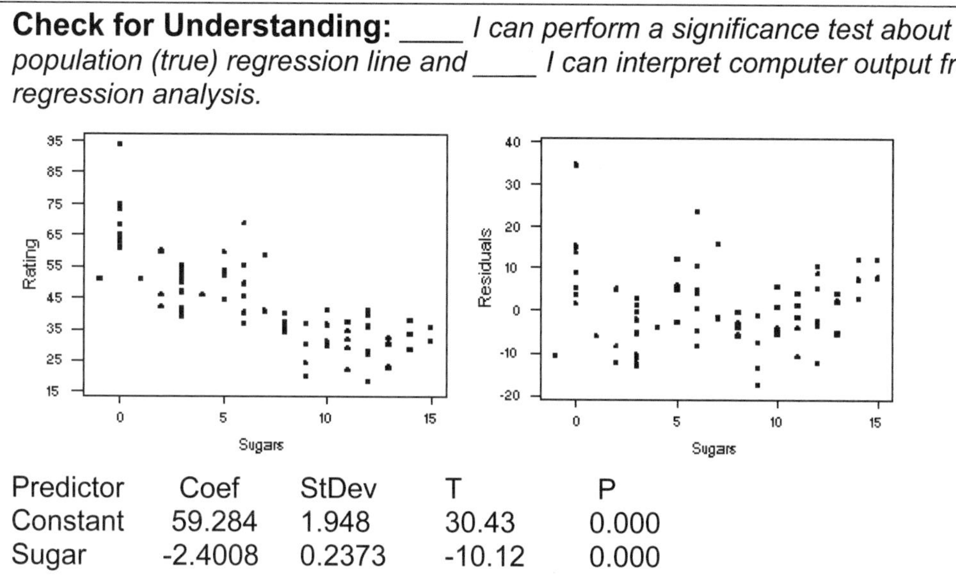


| Predictor | Coef | StDev | T | P |
|---|---|---|---|---|
| Constant | 59.284 | 1.948 | 30.43 | 0.000 |
| Sugar | -2.4008 | 0.2373 | -10.12 | 0.000 |

S = 9.196    R-Sq = 57.7%    R-Sq(adj) = 57.1%

The study in the previous Check for Understanding was expanded to include a total of 77 randomly selected cereals. The scatterplot, residual plot, and computer output of the regression analysis are noted above. Use this output to determine the LSRL for the sample data.

Interpret the slope in the context of the situation.

Is there convincing evidence that the slope of the true regression line is less than zero?

## Section 12.2: Transforming to Achieve Linearity

### Before You Read: Section Summary

You learned how to analyze linear relationships between two quantitative variables back in Chapter 3. In this section, you will learn how to deal with curved relationships. Since you already know how to model a linear relationship, you will learn how to transform data that show a curved relationship so that a linear model would be appropriate. That is, you will apply mathematical transformations to one or both variables to "straighten" out the scatterplot. By finding the linear model for the transformed data, you can make predictions involving the original data. The better the fit of your model, the better your prediction!

### "Where Am I Going?"

**Learning Targets:**

____ I can use transformations involving powers and roots to achieve linearity for a relationship between two quantitative variables.

____ I can use transformations involving logarithms to achieve linearity for a relationship between two quantitative variables.

____ I can determine which of several transformations does a better job of producing a linear relationship.

____ I can make predictions from a least-squares regression line involving transformed data.

### While You Read: Key Vocabulary and Concepts

transforming data:

power model:

logarithmic model:

exponential model:

### After You Read: "Where Am I Now?" Check for Understanding

### *Concept 1: Transforming with Powers and Roots*

When we know or suspect that a nonlinear relationship between two variables can be described by a model of the form $y = ax^p$, we have two strategies to transform the data to achieve linearity:

1) Raise all of the x values to the p power and plot $(x^p, y)$
2) Take the $p^{th}$ root of the $y$ values and plot $(x, \sqrt[p]{y})$.

We can then determine the least-squares regression line for the transformed data and use this equation to make predictions about the original data.

**Check for Understanding:** ____ *I can use transformations involving powers and roots to achieve linearity for a relationship between two quantitative variables.*

The following data represent the lengths (mm) and diameters (mm) of the humerus bones of the *Moleskius Primateum* species of monkeys once thought to inhabit Northern Minnesota.

| Diameter | 17.6 | 26 | 31.9 | 38.9 | 45.8 | 51.2 | 58.1 | 64.7 | 66.7 | 80.8 | 82.9 |
|---|---|---|---|---|---|---|---|---|---|---|---|
| Length | 159.9 | 206.9 | 236.8 | 269.9 | 300.6 | 323.6 | 351.7 | 377.6 | 384.1 | 437.2 | 444.7 |

Previous studies suggest the diameter and length are related by a power model of the form *length* = $a(diameter)^{0.7}$. Transform the original data and use least-squares regression to find an appropriate model for the transformed data.

You discover a portion of a *Moleskius Primateum* humerus bone with a diameter of 47mm. Use your model to predict how long the entire bone was.

### *Concept 2: Transforming with Logarithms*

In general, we don't know whether or not a power model is appropriate for describing the relationship between two quantitative variables. Some curved relationships are better summarized using an exponential model. A more efficient method for linearizing curved scatterplots involves using logarithms. To determine an appropriate model, use the following process:

1) Use logarithms to transform your data.
    - Plot (x, log y)
    - Plot (log x, log y)
2) Determine which transformation is most linear.
    - If (x, log y) is most linear, an exponential model may best describe (x, y)

- If (log x, log y) is most linear, a power model may best describe (x, y)

3) Find the appropriate linear model.
   - If (x, log y) is most linear, find the LSRL of the transformed data $\widehat{\log y} = a + bx$
   - If (log x, log y) is most linear, find the LSRL of the transformed data
     $\widehat{\log y} = a + b(\log [\![x)]\!]$

4) Use your model to make predictions for the original data.

**Check for Understanding:** ____ *I can use transformations involving logarithms to achieve linearity for a relationship between two quantitative variables,* ____ *I can determine which of several transformations does a better job of producing a linear relationship, and* ____ *I can make predictions from a least-squares regression line involving transformed data.*

The following data describe the number of police officers (thousands) and the violent crime rate (per 100,000 population) in a sample of states. Use these data to determine a model for predicting violent crime rate based on number of police officers employed. Show all appropriate plots and work.

| Police | 86.2 | 9.2 | 45 | 39.9 | 6 | 11.8 | 2.9 | 14.6 | 30.5 | 12.3 | 46.2 | 15.2 | 10.9 |
|---|---|---|---|---|---|---|---|---|---|---|---|---|---|
| Crime | 1090 | 559 | 1184 | 1039 | 303 | 951 | 132 | 763 | 635 | 726 | 840 | 373 | 523 |

Use your model to predict the violent crime rate for a state with 25,400 police officers.

## Chapter Summary: More About Regression

In this chapter, you learned how to apply your knowledge of inference to linear relationships. When you find a least-squares regression line for a set of sample data, you are constructing a model that approximates the true relationship between $x$ and $y$. By considering the sampling distribution of $b$, you can construct a confidence interval for the slope of the true regression line as well as test claims about the slope. In the event the relationship between two variables is curved, you can use transformations to "straighten" the scatterplot. You can find the least-squares regression line for the transformed data and make better predictions for the original relationship. The most common methods for transforming data involve taking powers, roots, or logarithms of one or both variables. To determine which transformation does a better job of "straightening" the relationship, examine residual plots.

### After You Read: "How Can I Close the Gap?"

Complete the vocabulary puzzle, multiple choice questions, and FRAPPY. Check your answers and your performance on each of the targets.

| Target | Got It! | Almost There | Needs Some Work |
|---|---|---|---|
| I can check conditions for performing inference about the slope of the population (true) regression line. | | | |
| I can interpret computer output from a least-squares regression analysis. | | | |
| I can construct and interpret a confidence interval for the slope of a regression line. | | | |
| I can perform a significance test about the slope of a population (true) regression line. | | | |
| I can use transformations involving powers and roots to achieve linearity for a relationship between two quantitative variables. | | | |
| I can use transformations involving logarithms to achieve linearity for a relationship between two quantitative variables. | | | |
| I can determine which of several transformations does a better job of producing a linear relationship. | | | |
| I can make predictions from a least-squares regression line involving transformed data. | | | |

Did you check "Needs Some Work" for any of the targets? If so, what will you do to address your needs for those targets?

*Learning Plan:*

# Chapter 12 Multiple Choice Practice

**Directions.** *Identify the choice that best completes the statement or answers the question. Check your answers and note your performance when you are finished.*

1. Is it possible to predict a student's GPA in their senior year from their GPA in the first marking period of their freshman year? A random sample of 15 seniors from the graduating class of 468 students is selected and both full-year GPA in their senior year ('Senior") and first-marking-period GPA in their freshman year ("Fresh") is recorded. A computer regression analysis and a residual plot for these data are given below.

| Predictor | Coef | SE Coef | T | P |
|---|---|---|---|---|
| Constant | 1.6310 | 0.5328 | 3.06 | 0.009 |
| Fresh | 0.5304 | 0.1789 | 2.96 | 0.011 |

S = 0.3558 R-Sq = 40.3% R-Sq(adj) = 35.7%

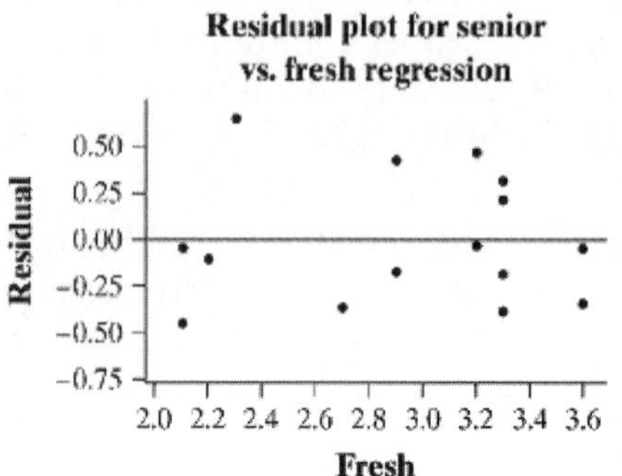


Which of the following is the estimate for the standard deviation of the sampling distribution of slopes?

| | |
|---|---|
| A. | 0.1789 |
| B. | 0.3558 |
| C. | 0.5304 |
| D. | 0.5328 |
| E. | 1.6310 |

2. The equation of the least-squares regression line is

| | |
|---|---|
| A. | $\widehat{Fresh} = 1.6310(Senior) + 0.5304$ |
| B. | $\widehat{Fresh} = 1.6310 + 0.5304(Senior)$ |
| C. | $\widehat{Senior} = 1.6310(Fresh) + 0.5304$ |
| D. | $\widehat{Senior} = 1.6310 + 0.5304(Fresh)$ |
| E. | $\widehat{Senior} = 0.5304(Fresh) + 0.1789$ |

3. Can we predict annual household electricity costs in a specific region from the number of rooms in the house? Below is a scatterplot of annual electricity costs (in dollars) *versus* number of rooms for 30 randomly-selected houses in Michigan, along with computer output for linear regression of electricity costs on number of rooms.

| Predictor | Coef | SE Coef | T | P |
|---|---|---|---|---|
| Constant | 406.9 | 164.8 | 2.47 | 0.020 |
| Rooms | 58.45 | 24.77 | 2.36 | 0.026 |

S = 246.735 R-Sq = 16.6% R-Sq(adj) = 13.6%

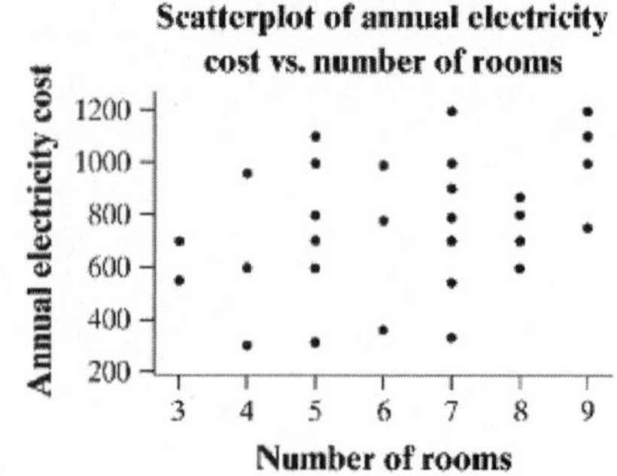


Assume the conditions for inference have been met. If we test the hypotheses $H_0$: $\beta = 0$ vs. $H_a$: $\beta > 0$ at the $\alpha = 0.05$ level. Which of the following is the appropriate conclusion?

| | |
|---|---|
| A. | Since the *P*-value of 0.020 is less than $\alpha$, we reject $H_0$. There is convincing evidence of a linear relationship between annual electricity costs and number of rooms in the population of Michigan homes. |
| B. | Since the *P*-value of 0.020 is greater than $\alpha$ we fail to reject $H_0$. We do not have enough evidence to conclude that there is a linear relationship between annual electricity costs and number of rooms |

| | in the population of Michigan homes. |
|---|---|
| C. | Since the *P*-value of 0.026 is greater than $\alpha$ , we accept $H_0$. We have convincing evidence that there is not a linear relationship between annual electricity costs and number of rooms in the population of Michigan homes. |
| D. | Since the *P*-value of 0.026 is less than $\alpha$ , we accept $H_0$. We have convincing evidence that there is not a linear relationship between annual electricity costs and number of rooms in the population of Michigan homes. |
| E. | Since the *P*-value of 0.026 is less than $\alpha$ , we reject $H_0$. We have convincing evidence of a linear relationship between annual electricity costs and number of rooms in the population of Michigan homes. |

4. Are high school students who like their English class more likely to enjoy their history class as well? Here is a residual plot for 30 randomly-selected students who were asked to rate how much they liked both English and history on a 0 to 5 scale (a higher rating means the student liked the subject more). [Data from 2004-5 Census at Schools survey in Canada.]

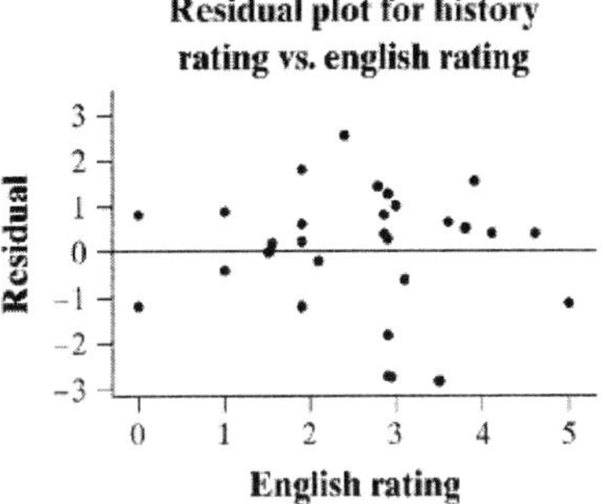


Which of the following conditions for inference does the residual plot suggest has not been satisfied?

| A. | The data come from a random sample. |
|---|---|
| B. | Observations for each student are independent. |
| C. | The variance of residuals is roughly equal for each value of English rating. |
| D. | For each value of English rating, the distribution of history rating is roughly Normal. |
| E. | Mean History rating is a linear function of English rating. |

5. Consider the output for the regression analysis on the situation from question 4.

| Predictor | Coef | SE Coef | T | P |
|---|---|---|---|---|
| Constant | 1.1867 | 0.5574 | 2.13 | 0.042 |
| English | 0.5254 | 0.1995 | 2.63 | 0.014 |

S = 1.37707 R-Sq = 19.8% R-Sq(adj) = 17.0%

Assume the conditions for regression inference have been satisfied. What does the quantity R-Sq = 19.8% represent?

| A. | The correlation of history rating and English rating—a measure of the strength of the linear relationship between the two variables. |
|---|---|
| B. | The average deviation of observed history ratings from the predicted history ratings, expressed as a percentage of the predicted history rating. |
| C. | The average deviation of observed English ratings from the predicted English ratings, expressed as a percentage of the predicted English rating. |
| D. | The percentage of variation in history rating that can be explained by the regression of history rating on English rating. |
| E. | The LSRL is accurately predicts history rating 19.8% of the time. |

6. Which of the following is the 95% confidence interval for the population slope?

| A. | 1.1867 ± 1.960(0.5574) |
|---|---|
| B. | 1.1867 ± 2.048(0.5574) |
| C. | 0.5254 ± 1.960(0.1995) |
| D. | 0.5254 ± 2.048(0.1995) |
| E. | 0.5254 ± 2.630(1.37707) |

7. Suppose we measure a response variable *Y* for several values of an explanatory variable *X*. A scatterplot of log *Y* versus log *X* looks approximately like a negatively-sloping straight line. We may conclude that

| | |
|---|---|
| A. | the rate of growth of *Y* is positive, but slowing down over time. |
| B. | an exponential growth model would approximately describe the relationship between *Y* and *X*. |
| C. | a power model would approximately describe the relationship between *Y* and *X*. |
| D. | the relationship between *Y* and *X* is a positively-sloping straight line. |
| E. | the residual plot of the regression of log *Y* on log *X* would have a "U-shaped" pattern suggesting a non-linear relationship. |

8. Suppose the relationship between a response variable *y* and an explanatory variable *x* is modeled well by the equation $y = 3.6(0.32)^x$. Which of the following plots is most likely to be roughly linear?

| | |
|---|---|
| A. | A plot of *y* against *x* |
| B. | A plot of *y* against log *x* |
| C. | A plot of log *y* against *x* |
| D. | A plot of $10^y$ against *x* |
| E. | A plot of log *y* against log *x* |

9. Use of the Internet worldwide increased steadily from 1990 to 2002. A scatterplot of this growth shows a strongly non-linear pattern. However, a scatterplot of *ln* Internet Users *versus* Year is much closer to linear. Below is a computer regression analysis of the transformed data (note that natural logarithms are used).

| Predictor | Coef | SE Coef | T | P |
|---|---|---|---|---|
| Constant | -951.10 | 43.45 | -21.89 | 0.000 |
| Year | 0.4785 | 0.02176 | 21.99 | 0.000 |

S = 0.2516 R-Sq = 98.2% R-Sq(adj) = 98.0%

Which of the following best describe the model that is given by this computer printout?

| | |
|---|---|
| A. | The linear model: $\widehat{users} = -951.10 + 0.4785(year)$ |
| B. | The power model: $\widehat{users} = e^{-951.10}(year)^{0.4785}$ |
| C. | The power model: $\widehat{users} = 10^{-951.10}(year)^{0.4785}$ |
| D. | The exponential model: $\widehat{users} = e^{-951.10}(e^{0.4785})^{year}$ |
| E. | The exponential model: $\widehat{users} = 10^{-951.10}(10^{0.4785})^{year}$ |

10. Like most animals, small marine crustaceans are not able to digest all the food they eat. Moreover, the percentage of food eaten that is assimilated (that is, digested) decreases as the amount of food eaten increases. A residual plot for the regression of Assimilation rate (as a percentage of food intake) on Food Intake (in g/day) is shown below.

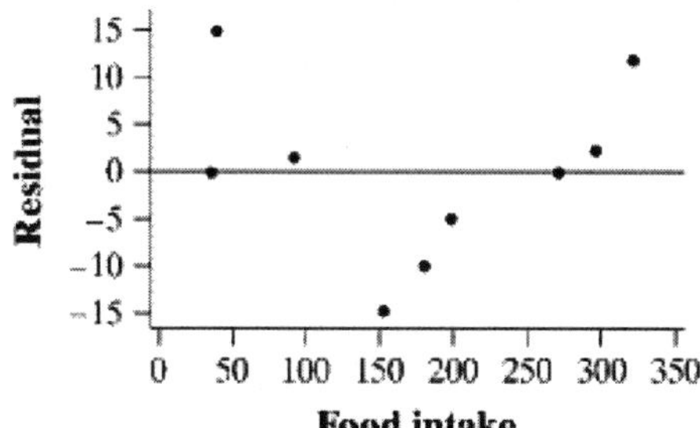


A scatterplot of ln Assimilation *versus* ln Food Intake is strongly linear, suggesting that a linear regression of these transformed variables may be more appropriate. Below is a computer regression analysis of the transformed data (note that natural logarithms are used).

| Predictor | Coef | SE Coef | T | P |
|---|---|---|---|---|
| Constant | 6.3324 | 0.5218 | 12.14 | 0.000 |
| ln Food Intake | -0.6513 | 0.1047 | -6.22 | 0.000 |

S = 0.247460 R-Sq = 84.7% R-Sq(adj) = 82.5%

When food intake is 250 g/day , what is the predicted assimilation rate from this model?

| | |
|---|---|
| A. | 2.7% |
| B. | 15.4% |
| C. | 27.4% |
| D. | 34.3% |
| E. | 54.4% |

## Multiple Choice Answers

| Problem | Answer | Concept | Right | Wrong | Simple Mistake? | Need to Study More |
|---|---|---|---|---|---|---|
| 1 | A | SE of Slope from Output | | | | |
| 2 | D | Regression Equation from Output | | | | |
| 3 | E | Significance Test from Output | | | | |
| 4 | C | Conditions for Inference | | | | |
| 5 | D | Confidence Interval for Slope | | | | |
| 6 | B | Conclusion from Significance Test | | | | |
| 7 | C | Interpreting log-log Scatterplot | | | | |
| 8 | C | Exponential Functions/Transfromations | | | | |
| 9 | D | Semi-log Transformation | | | | |
| 10 | B | Prediction from log-log Transformation | | | | |

## FRAPPY! Free Response AP Problem, Yay!

The following problem is modeled after actual Advanced Placement Statistics free response questions. Your task is to generate a complete, concise response in 15 minutes. After you generate your response, view two example solutions and determine whether you feel they are "complete," "substantial," "developing" or "minimal." If they are not "complete," what would you suggest to the student who wrote them to increase their score? Finally, you will be provided with a rubric. Score your response and note what, if anything, you would do differently to increase your own score.

Paul is interested in purchasing a digital camera and notices that as each model's image quality (in megapixels) increases, the cost appears to increase linearly. A scatterplot and regression output of the megapixels vs. cost for seven randomly chosen camera models is below:

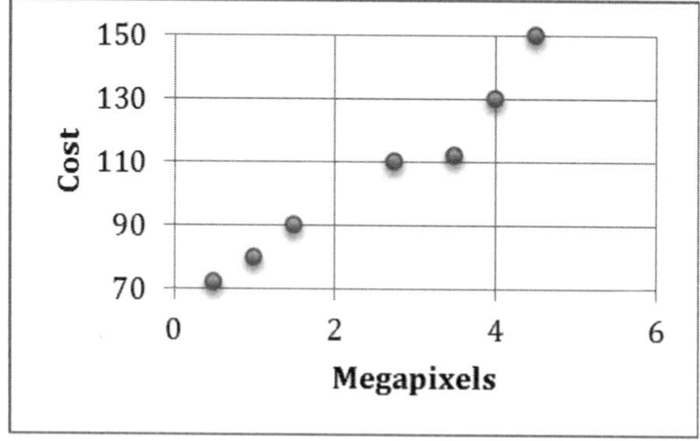


Regression Analysis: Cost versus Megapixels

| Predictor | Coef | SE Coef | T | P |
|---|---|---|---|---|
| Constant | 63.457 | 2.387 | 26.58 | 0.000 |
| Speed | 16.2809 | 0.8192 | 19.88 | 0.000 |

S = 3.087 R-Sq = 98.7% R-Sq (adj) = 98.5%

a) Using the regression output, write the equation of the fitted regression line in context.

b) Interpret the slope and y-intercept in the context of the problem.

c) Construct and interpret a 98% confidence interval for the slope. Assume the conditions for inference have been met.

**Student Response 1:**

a) y = 63.457 + 16.2809x

b) The intercept of 63.457 means a camera with 0 megapixels will cost about 63 dollars. This doesn't really make any sense. The slope of 16.28 means for each increase of one megapixel, the cost goes up $16.28.

c) 98% CI for slope: $b \pm t^* SE_b$

$16.2809 \pm 3.365\ (0.8192) = (13.52, 19.03)$

We are 98% confident the true slope falls between 13.52 and 19.03.

How would you score this response? Is it substantial? Complete? Developing? Minimal? Is there anything this student could do to earn a better score?

**Student Response 2:**

a) $\widehat{cost} = 63.453 + 16.2809(megapixels)$

b) The intercept doesn't make sense in terms of the problem as it states a camera with NO megapixels would be predicted to cost $63.46. The slope tells us that for each increase of 1 megapixel in picture quality, we predict the cost will increase by approximately $16.28 on average.

c) Since the conditions are met, we can construct a 98% t-interval for the true slope.

$16.2809 \pm 3.365\ (0.8192) = (13.52, 19.03)$

We are 98% confident the interval from 13.52 to 19.03 captures the true slope of the relationship between megapixels and cost. That is, we are 98% confident for each increase of one megapixel, the predicted cost of a camera will increase between $13.52 and $19.03.

How would you score this response? Is it substantial? Complete? Developing? Minimal? Is there anything this student could do to earn a better score?

## Scoring Rubric

Use the following rubric to score your response. Each part receives a score of "Essentially Correct," "Partially Correct," or "Incorrect." When you have scored your response, reflect on your understanding of the concepts addressed in this problem. If necessary, note what you would do differently on future questions like this to increase your score.

### Intent of the Question

The primary goals of this question are to assess your ability to (1) interpret standard computer output; (2) interpret a linear model in context; (3) construct and interpret a confidence interval for the slope of a regression line;

### Solution

a) $\widehat{cost} = 63.453 + 16.2809(megapixels)$

b) Intercept = 63.457. This is the predicted cost when megapixels = 0.
Slope = 16.2809. We predict a cost increase of $16.28 for each increase of 1 megapixel.

c) 98% Confidence Interval: 16.2809 ± 3.365(0.8192) = (13.52, 19.03). We are 98% confident the interval from 13.52 to 19.03 captures the true slope of the regression line for megapixels and cost.

### Scoring:

Each element scored as essentially correct (E), partially correct (P), or incorrect (I).

**a)** Essentially correct if the equation is properly stated and variables are defined in the context of the problem. Partially correct if the equation is properly stated, but the variables are not defined or left as x and y.

**b)** Essentially correct if both the intercept and slope are identified and interpreted correctly. Partially correct if only one is properly identified and interpreted. Note, the slope must indicate a *predicted* increase.

**c)** Essentially correct if a correct interval is constructed (with formula shown) and interpreted in the context of the problem. Partially correct if supporting work is not provided or if the interpretation is incorrect or lacks context.

**4 Complete Response**
All three parts essentially correct

**3 Substantial Response**
Two parts essentially correct and one part partially correct

**2 Developing Response**
Two parts essentially correct and no parts partially correct
One part essentially correct and two parts partially correct
Three parts partially correct

**1 Minimal Response**
One part essentially correct and one part partially correct
One part essentially correct and no parts partially correct
No parts essentially correct and two parts partially correct

# Chapter 12: More About Regression

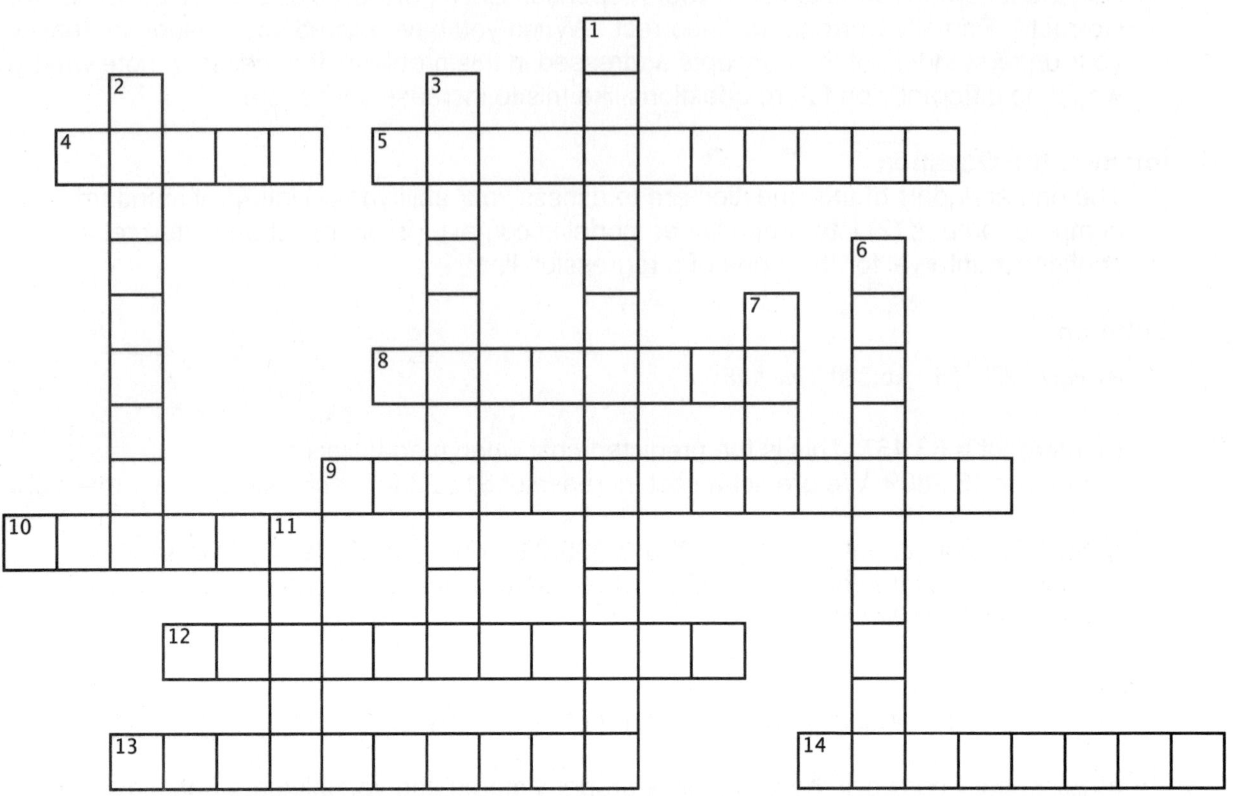


**Across**

4. if (ln(x), ln(y)) is linear, a/an ____ model may be most appropriate for (x,y)
5. another name for the x-variable
8. another name for the y variable
9. the coefficient of ____ indicates the fraction of variability in predicted y values that can be explained by the LSRL of y on x
10. when we construct a LSRL for a set of observed data, we construct a ____ regression line
12. the _____ coefficient is a measure of the strength of the linear relationship between two quantitative variables
13. the line that describes the relationship between an explanatory and response variable
14. observed y - predicted y

**Down**

1. when data displays a curved relationship, we can perform a ____ to "straighten" it out
2. the common mathematical transformation use in this chapter to achieve linearity for a relationship between two quantitative variable
3. if (x, ln(y)) is linear, a/an ____ model may be most appropriate for (x,y)
6. to estimate the population (true) slope, we ca construct a ____ interval
7. the Greek letter we use to represent the slope the population (true) regression line
11. to calculate a confidence interval, we must us the standard ____ of the slope

# Solutions

# Chapter 1: Solutions

**Introduction Concept 2:** ____ *I can identify key characteristics of a set of data.*
1) This data set describes the students in Mr. Buckley's class (James, Jen, ..., Sharon)

2) Quantitative: ACT Score, GPA          Categorical: Gender, Favorite Subject

3) The ACT scores range from 28 to 35 with a center around 32. Four of the scores are "bunched up" in the 32-35 range, while Jonathan's score of 28 doesn't quite fit with the rest of the scores.

4) It appears the students who prefer math and science had higher scores on the ACT. However, we can not infer there is a large difference between the scores of these students and those who do not prefer math and science.

**Section 1.1 Concept 1:** ____ *I can display categorical data with pie charts or bar graphs.*
1) It appears the majority of individuals prefer Goodbye Blue Monday and One Mean Bean. Very few people prefer the national chain.

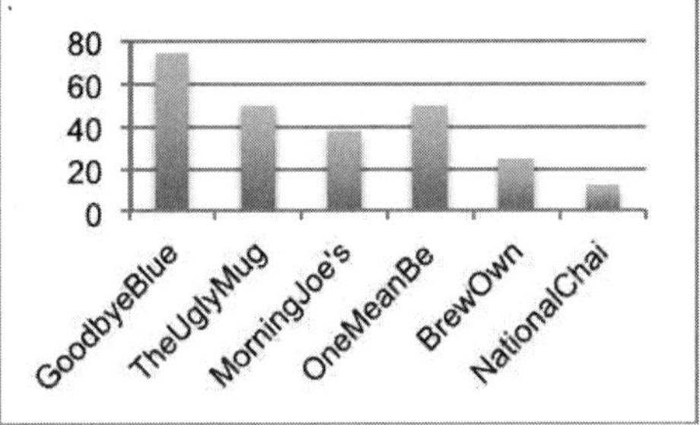


2) The pie chart does not just show actual counts. Rather, it displays the percent of the total that preferred each coffee-shop. It is clear, again, from this chart that the overwhelming preference is for Goodbye Blue Monday, One Mean Bean, and The Ugly Mug with 70% of residents polled preferring them.

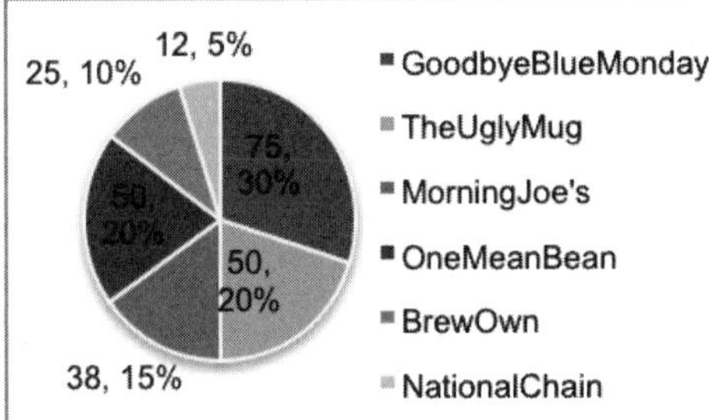


**Section 1.1 Concept 2:** ____ *I can describe the relationship between two categorical variables.*
State: What is the relationship between gender and coffee preference?
Plan: We suspect gender might influence coffee preference, so we'll compare the conditional distributions of coffee preference for men alone and women alone.

| Preference | Male | Female |
|---|---|---|
| National Chain | 17.8% | 14.0% |
| One Mean Bean | 2.8% | 18.3% |
| The Ugly Mug | 27.1% | 5.4% |
| Goodbye Blue Monday | 31.8% | 19.4% |
| Home-brewed 18.7% | | 34.4% |
| Don't drink coffee | 1.9% | 8.6% |

Do: We'll make a side-by-side bar graph to compare the preferences of males and females.
Conclude: Based on the sample, it appears men prefer to get their coffee from The Ugly Mug and Goodbye Blue Monday while women are more likely (34.4% vs 18.7%) to home brew their coffee.

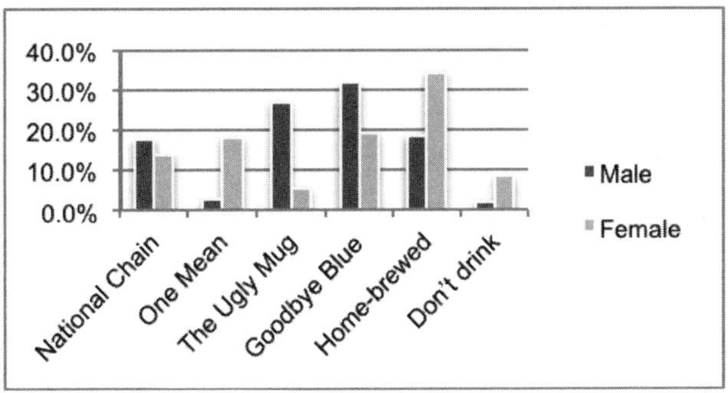

**Section 1.2 Concept 1:** ____ *I can construct and interpret dotplots and stemplots.*
1) The distribution of gas mileage appears to be fairly symmetric, centered at about 37 mpg, and ranging from 31.8 to 41 mpg. There do not appear to be any extreme values.

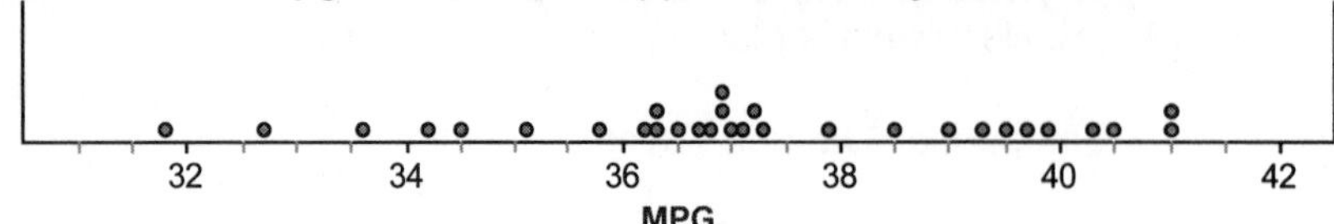


2) Stemplot:

31 | 8
32 | 7
33 | 6
34 | 2 5
35 | 1
36 | 2 3 3 5 7 8 9 9
37 | 0 1 2 3 9
38 | 5
39 | 0 3 5 7 9
40 | 3 5
41 | 0 0

**Section 1.2 Concept 2:** ____ *I can construct and interpret a histogram.*
(a) Histograms may vary depending on "bin widths" and starting values. The distribution of mileage for the 50 cars is mound shaped, with a slight skew to the right. The center of the distribution is approximately 37mpg with a range from 31 to 42. There do not appear to be any extreme values.
(b) To get your calculator to match the histogram you constructed by hand you will need to adjust the Ymin and Yscl in window settings.

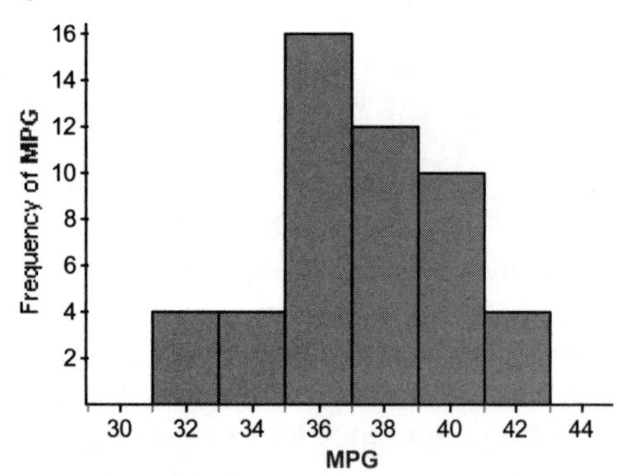


**Section 1.3 Concept 2:** ____ *I can calculate and interpret measures of center.*
1) The plot suggests the distribution is slightly skewed to the right. Therefore, the mean will probably be greater than the median since it tends to get pulled in the direction of the tail.

2) The average amount of time necessary to complete the logic puzzle was 32.67 seconds.

3) Median = 29. This is the "middle" time in the distribution. Half of the students took longer than 29 seconds to complete the puzzle and half took less than 29 seconds.

4) Since there is some skewness, the median would be a better measure as it is not affected by the tail or extreme values. It is a more accurate measure of center for this distribution.

**Section 1.3 Concept 3:** ____ *I can calculate and interpret measures of spread.*
(a) The distribution of book lengths (in pages) is fairly symmetric centered at about 330 with outliers at the minimum of 170 and 242, Q1 = 314, Median = 330, Q3 = 344, and maximum of 374.

(b) These data have a mean of 316.27 pages and standard deviation 51.63 pages. Anne's favorite books tend to be within about 52 pages of 316, on average.

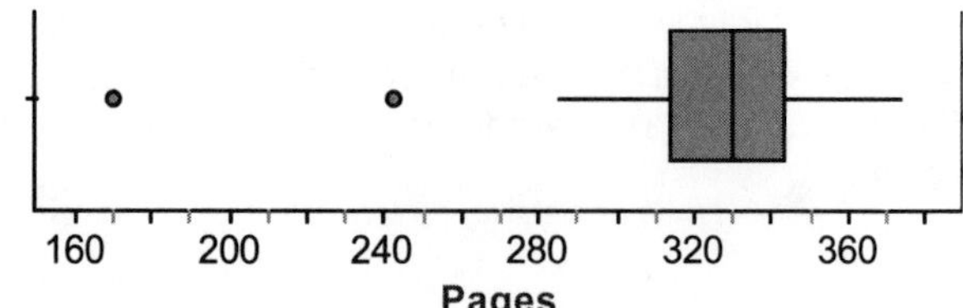

# Chapter 1: Exploring Data

**Across**

3. The average distance of observations from their mean (two words)
5. The average squared distance of the observations from their mean
9. Displays the counts or percents of categories in a categorical variable through differing heights of bars
12. Tells you what values a variable takes and how often it takes these values
13. Displays a categorical variable using slices sized by the counts or percents for the categories
16. When specific values of one variable tend to occur in common with specific values of another
18. A measure of center, also called the average
19. A graphical display of quantitative data that involves splitting the individual values into two components
21. One of the simplest graphs to construct when dealing with a small set of quantitative data
22. Drawing conclusions beyond the data at hand
23. The shape of a distribution if one side of the graph is much longer than the other
24. What we call a measure that is relatively unaffected by extreme observations

**Down**

1. The objects described by a set of data
2. The midpoint of a distribution of quantitative data
4. A ____ distribution describes the distribution of values of a categorical variable among individuals who have a specific value of another variable.
6. A variable that places an individual into one of several groups or categories
7. A characteristic of an individual that can take different values for different individuals
8. When comparing two categorical variables, we can orgainze the data in a ___-___ ____.
9. A graphical display of the five-number summary
10. A graphical display of quantitative data that shows the frequency of values in intervals by using bars
11. A variable that takes numerical values for which it makes sense to find an average
14. The shape of a distribution whose right and left sides are approximate mirror images of each other
15. These values lie one-quarter, one-half, and three-quarters of the way up the list of quantitative data
17. A value that is at least 1.5 IQRs above the third quartile or below the first quartile
20. When exploring data, don't forget your ____

# Chapter 2: Solutions

**Section 2.1 Concept 1:** ____ *I can describe the location of an observation in a distribution using percentiles and* ____ *I can interpret cumulative relative frequency graphs.*

1) James earned a score of 33. His score falls above 17 of the 23 students. Approximately 78% of students scored at or below 33.

2) Heather earned a score of 12. Her score falls above 3 of the 23 students. Approximately 17% of students scored at or below 12.

3) Using the ogive, a score of 38 is at about the 80$^{th}$ percentile.

4) Using the ogive, Q1 is about 14, the median is about 25, and Q3 is about 30.

**Section 2.1 Concept 2:** ____ *I can describe the location of an observation in a distribution using z-scores.*

1) Mean = 25.04 Standard Deviation = 14.20

2) $z = \frac{45 - 25.04}{14.2} = 1.41$ Paul's score falls 1.41 standard deviations above the mean score.

3) $z = \frac{2 - 25.04}{14.2} = -1.62$ Carl's score falls 1.62 standard deviations below the mean score.

4) First quiz: $z = \frac{45 - 25.04}{14.2} = 1.41$ Next quiz: $z = \frac{47 - 32}{6.5} = 2.31$

Paul's z-score is greater for the next quiz. He scored better relative to the mean on this quiz.

**Section 2.1 Concept 4:** ____ *I can describe the density curve for a distribution of data.*

1) The total area under the curve is 1.

2) 12% of scores fell between 28 and 40?

3) Since this curve is skewed left, the mean will be less than the median. The median probably falls around 50, while the mean is likely to be between 40 and 50. It is difficult to tell without the actual values, though.

**Section 2.2 Concept 2:** ____ *I can perform Normal distribution calculations and interpret their results.*

1) 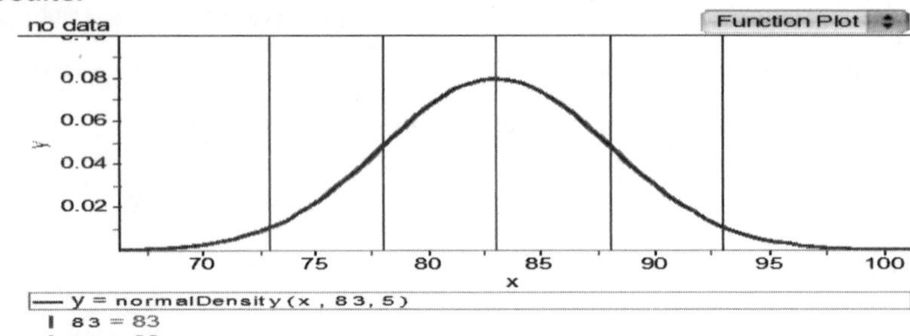

2) P(78<x<83) = 0.34   P(83<x<93) = .475   P(78<x<93) = 0.34 + 0.475 = 81.5%

3) $z = \frac{90-83}{5} = 1.4$ P(z>1.40) = 1 – 0.9192 = 0.0808. 8.08% of quizzes earn an "A".

4) $z = \frac{71-83}{5} = -2.4$   $z = \frac{95-83}{5} = 2.4$ -> 0.9918 – 0.0082 = 0.9836. 98.36% of scores would fall between 71 and 95.

5) The 20[th] percentile corresponds to a *z*-score of about -0.84. x = 83 - 0.84(5) = 78.8.

**Section 2.2 Concept 3:** ____ *I can justify whether or not a distribution can be considered Normal.*

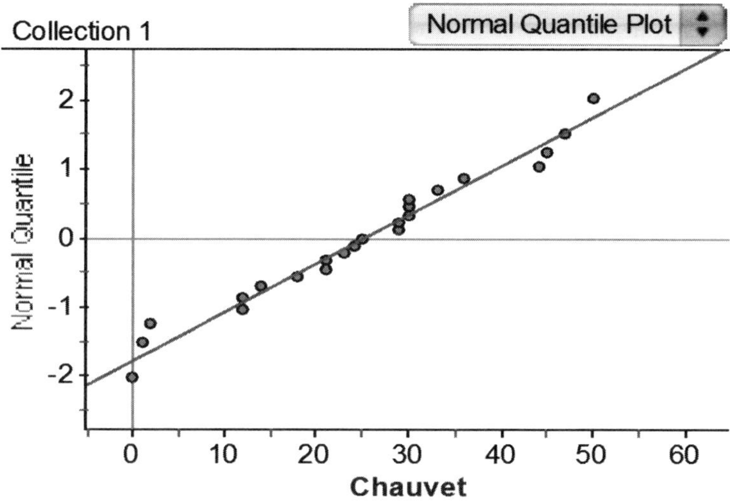


The quickest way to assess Normality is to construct a Normal quantile plot. If the plot displays a linear pattern, the data can be assumed to be approximately Normal. It appears the quiz scores for Ms. Chauvet's class are approximately Normal. You could also justify this by counting how many scores fall within one-, two-, and three- standard deviations of the mean and determining whether or not that reflects the 68-95-99.7 rule. Keep in mind, you will rarely find a distribution that is perfectly Normal!

# Chapter 2: Modeling Distributions of Data

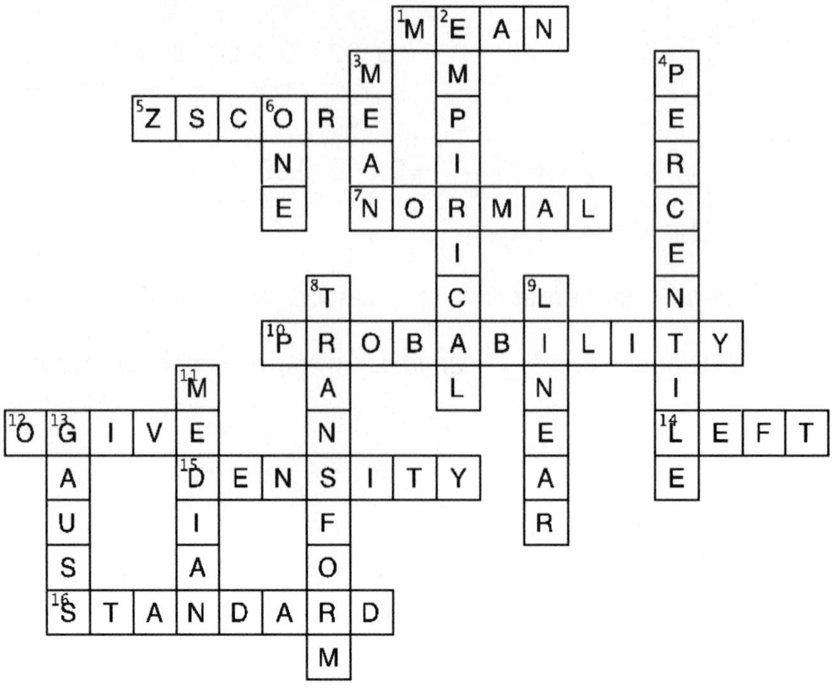

**Across**

1. The balance point of a density curve, if it were made of solid material [MEAN]
5. The standardized value of an observation [ZSCORE]
7. These common density curves are symmetric and bell-shaped [NORMAL]
10. A Normal ______ plot provides a good assessment of whether a data set is approximately Normally distributed [PROBABILITY]
12. Another name for a cumulative relative frequency graph [OGIVE]
14. The standard Normal table tells us the area under the standard Normal curve to the ___ of z [LEFT]
15. A ___ curve is a smooth curve that can be used to model a distribution [DENSITY]
16. This Normal distribution has mean 0 and standard deviation 1 [STANDARD]

**Down**

2. The ____ rule is also known as the 68-95-99.7 rule for Normal distributions [EMPIRICAL]
3. To standardize a value, subtract the and divide by the standard deviation [MEAN]
4. The value with p percent of the observations less than it [PERCENTILE]
6. The area under any density curve is always equal to [ONE]
8. We ___ data when we change each value by adding a constant and/or multiplying by a constant. [TRANSFORM]
9. If a Normal probability plot shows a _____ pattern, the data are approximately Normal [LINEAR]
11. The point that divides the area under a density curve in half [MEDIAN]
13. This mathematician first applied Normal curves to data to errors made by astronomers and surveyors [GAUSS]

# Chapter 3: Solutions

**Section 3.1 Concept 1:** ____ *I can identify explanatory and response variables.*

Explanatory: test anxiety Response: performance on the test
Explanatory: volume of hippocampus Response: verbal retention

**Section 3.1 Concept 2:** ____ *I can construct and interpret a scatterplot.* ____ *I can describe the form, direction, and strength of a relationship displayed in a scatterplot.*

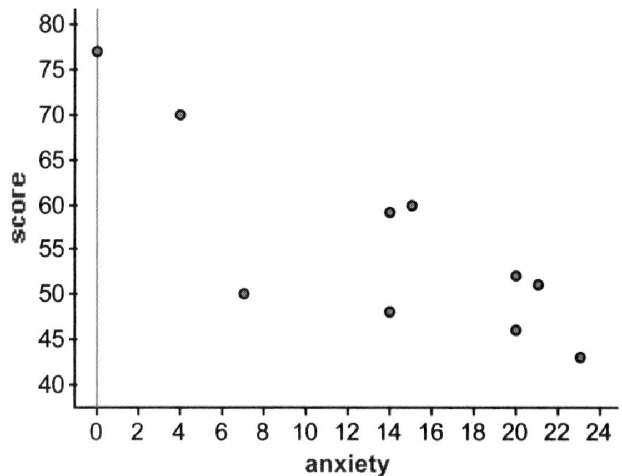


The scatterplot suggests a moderate, negative, linear relationship between anxiety and exam performance. As anxiety increases, exam performance decreases.

**Section 3.1 Concept 3:** ____ *I can calculate and interpret correlation.*

$r$ = -0.8185 There is a strong, negative, linear relationship between anxiety and test score.

**Section 3.2 Concept 2:** ____ *I can construct, interpret, and apply the least-squares regression line.*

a) predicted score = 71.3 -1.14(anxiety)

b) slope = -1.14. For each increase of one unit in anxiety scores, we predict about a 1.14 point decrease in exam performance.

c) predicted score = 71.3 -1.14(15) = 54.2.

d) residual = observed – predicted = 60 – 54.2 = 5.8.

e) No, 35 is far greater than the maximum observed anxiety score. Extrapolation such as this could result in a poor prediction as we do not know if the linear model applies beyond the observed data. (for all we know, the exam scores could actually start to increase!)

**Section 3.3 Concept 3:** ____ *I can assess how well the least-squares regression line fits the data.*

1) The residual plot does not show any obvious pattern. It appears the linear model may be appropriate for making predictions of exam scores.

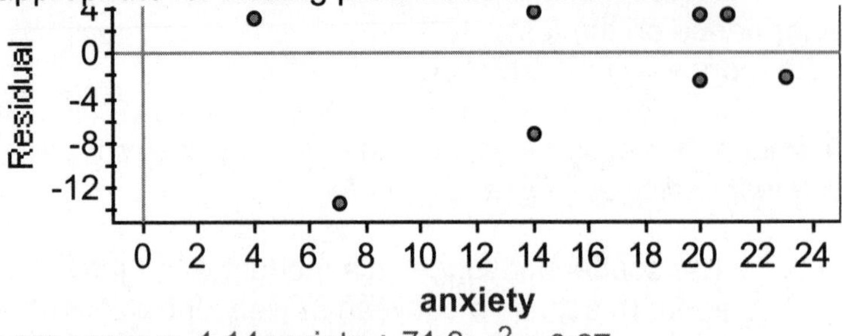


— score = -1.14anxiety + 71.3; $r^2$ = 0.67

2) $r = 0.8185$ $r^2$=0.67 There is a strong, negative, linear relationship between anxiety score and exam score. Approximately 67% of the variability in predicted exam scores can be explained by the least squares regression model of anxiety score on exam score.

**Section 3.3 Concept 4:** ____ *I can construct or identify the equation of a least squares regression line.*

1) predicted algae level = 42.8477 + 0.4762(temperature)

2) Slope = 0.4762. For each increase of 1 degree F, we predict an increase of 0.4762 parts per million in algae level.

3) $r = \sqrt{0.917} = 0.957$. There is a strong, positive, linear relationship between the temperature and algae level.

4) s = 0.4224. The predicted algae levels differ from the observed levels by an average of about 0.4224 units.

## Chapter 3: Describing Relationships

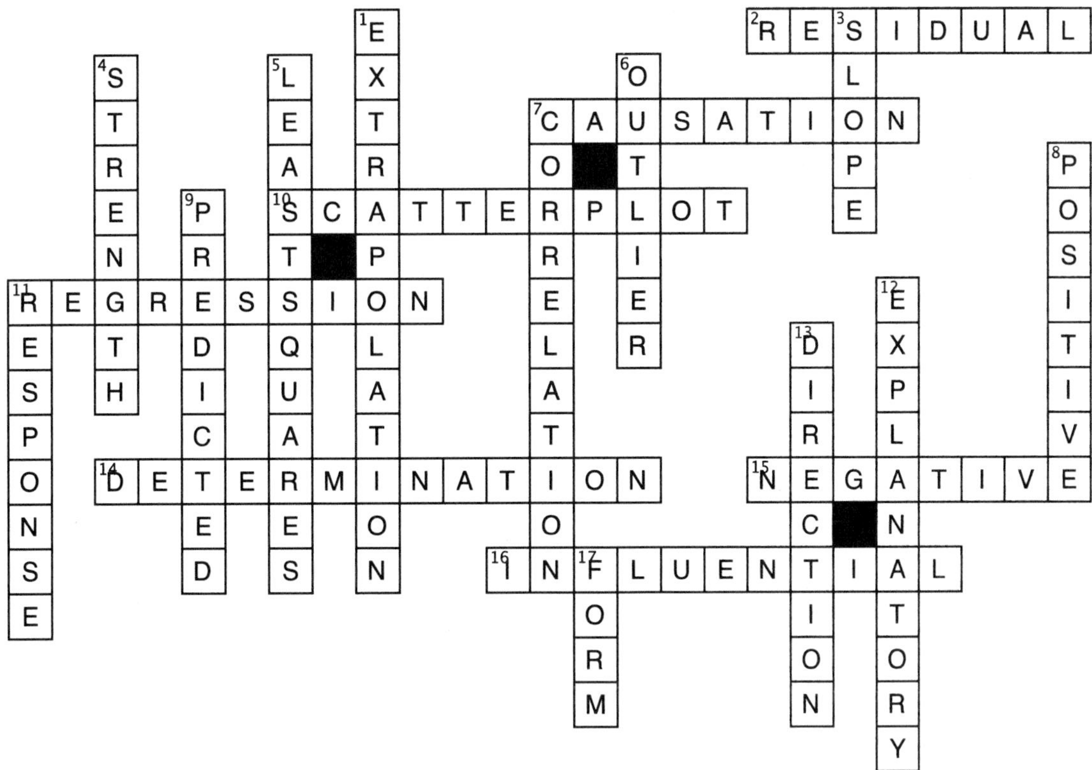

**Across**

2. the difference between an observed value of the response and the value predicted by a regression line [RESIDUAL]
7. Important note: Association does not imply ______. [CAUSATION]
10. graphical display of the relationship between two quantitative variables [SCATTERPLOT]
11. line that describes the relationship between two quantitative variables [REGRESSION]
14. the coefficient of _______ describes the fraction of variability in y values that is explained by least squares regression on x. [DETERMINATION]
15. A _____ association is defined when above average values of one variable are accompanied by below average values of the other. [NEGATIVE]
16. individual points that substantially change the correlation or slope of the regression line [INFLUENTIAL]

**Down**

1. the use of a regression line to make a prediction far outside the observed x values [EXTRAPOLATION]
3. the amount by which y is predicted to change when x increases by one unit [SLOPE]
4. The ______ of a relationship in a scatterplot is determined by how closely the point follow a clear form. [STRENGTH]
5. the ____-_____ regression line is also known as the line of best fit (2 words) [LEASTSQUARES]
6. an individual value that falls outside the overall pattern of the relationship [OUTLIER]
7. value that measures the strength of the linear relationship between two quantitative variables [CORRELATION]
8. A ______ association is defined when above average values of the explanatory are accompanied by above average values of the response [POSITIVE]
9. y-hat is the _______ value of the y-variable for a given x [PREDICTED]
11. variable that measures the outcome of a study [RESPONSE]
12. variable that may help explain or influence changes in another variable [EXPLANATORY]
13. The _____ of a scatterplot indicates a positive or negative association between the variables. [DIRECTION]
17. The ____ of a scatterplot is usually linear or nonlinear. [FORM]

# Chapter 4: Solutions

**Section 4.1 Concept 1:** ____ *I can identify the population and sample in a sampling situation.*

The population is all sentences and words used in the popular Algebra 1 textbooks. The sample consists of the sentences and words in the 10 randomly selected paragraphs.

**Section 4.1 Concept 2:** ____*I can describe how to use a table of random numbers or a random number generator to select a simple random sample.*

Assign each student a two-digit label 01-23. Read two-digit blocks across the random number table until 4 of the labels are selected, ignoring repeats and labels from 24-00.
The four selected are 19: Rohnkol, 22: Wilcock, 05: Buckley, 13: Lacey

**Section 4.2 Concept 1:** ____*I can explain how lurking variables can lead to confounding*

1. This is an observational study since no treatment was imposed on the subjects. Test scores and shoe size were observed. No effort was made to influence either variable.
2. The explanatory variable would be the shoe size and response would be test score.
3. One possible confounding variable could be age of the students. Older students would have bigger feet, on average, and would most likely have higher test scores.

**Section 4.2 Concept 2:** ____*I can identify experimental units, explanatory variables, treatments, and response variables in an experiment*
____ *I can describe a completely randomized design for an experiment*

1. Experimental units: Mr. Tyson's students  Explanatory variable: listening to classical music (yes or no)  Response variable: test scores
2. A potential lurking variable could be an existing musical preference based on test score. Perhaps higher performing students prefer listening classical music. Maybe students who listen to classical music have more resources/opportunities to score well than those who don't.
3. The 150 students should be randomly assigned to two groups. This could be accomplished by drawing names from a hat until 75 are in one group and 75 are in another. Both groups will receive the same instruction from Mr. Tyson. However, one group will be assigned to listen to classical music while studying and the other group will study in silence. All students will take the same assessment and their average results will be compared.

**Section 4.2 Concept 3:** ____ *I can distinguish between completely randomized designs and block designs*

(example) Suppose Mr. Tyson suspects students enrolled in a music class may have higher scores than those who don't. In order to ensure not all students enrolled in a music class are assigned to listen to classical music (which could happen through random assignment), he should block by enrollment. He should block all students enrolled in a music class together and all students who are not enrolled in music should be blocked separately. Then, he should randomly assign half of the students in each block to listen to classical music while studying and the other half should study in silence. Then all students should take the same assessment and the results within each block should be compared.

# Chapter 4: Designing Studies

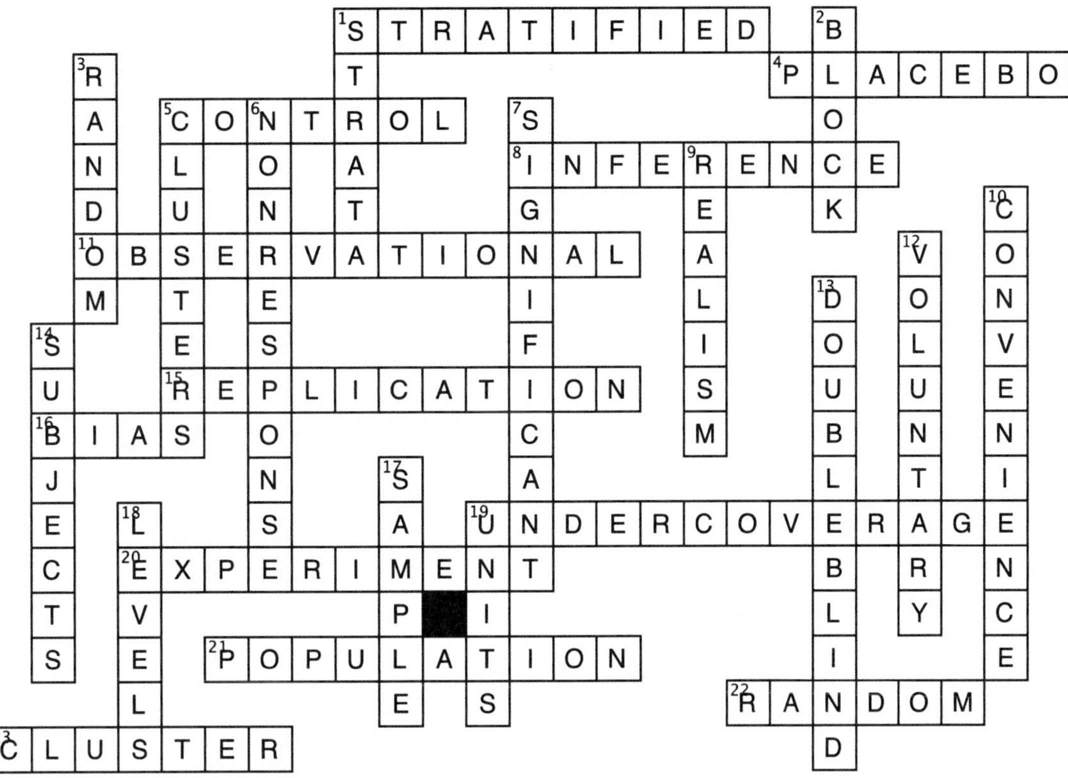

**Across**

1. a _____ random sample consists of separate simple random samples drawn from groups of similar individuals [STRATIFIED]
4. a "fake" treatment that is sometimes used in experiments [PLACEBO]
5. the effort to minimize variability in the way experimental units are obtained and treated [CONTROL]
8. the process of drawing a conclusion about the population based on a sample [INFERENCE]
11. this type of student can not be used to establish cause-effect relationships [OBSERVATIONAL]
15. the practice of using enough subjects in an experiment to reduce chance variation [REPLICATION]
16. a study that systematically favors certain outcomes shows this [BIAS]
19. this occurs when some groups in the population are left out of the process of choosing the sample [UNDERCOVERAGE]
20. a study in which a treatment is imposed in order to observe a reasponse [EXPERIMENT]
21. the entire group of individuals about which we want information [POPULATION]
22. a simple _____ sample consists of individuals from the population, each of which has an equally likely chance of being chosen [RANDOM]
23. a _____ sample consists of a simple random sample of small groups from a population [CLUSTER]

**Down**

1. groups of similar individuals in a population [STRATA]
2. a group of experimental units that are similar in some way that may affect the response to the treatments [BLOCK]
3. the rule used to assign experimental units to treatments is ____ assignment [RANDOM]
5. smaller groups of individuals who mirror the population [CLUSTERS]
6. this occurs when an individual chosen for the sample can't be contacted or refuses to participate [NONRESPONSE]
7. an observed effect that is too large to have occurred by chance alone [SIGNIFICANT]
9. a lack of ____ in an experiment can prevent us from generalizing the results [REALISM]
10. a sample in which we choose individuals who are easiest to reach [CONVENIENCE]
12. a ____ response sample consists of people who choose themselves by responding to a general appeal [VOLUNTARY]
13. neither the subjects nor those measuring the response know which treatment a subject received (two words) [DOUBLEBLIND]
14. when units are humans, they are called [SUBJECTS]
17. the part of the population from which we actually collect information [SAMPLE]
18. another name for treatments [LEVELS]
19. the individuals on which an experiement is done are experimental ____ [UNITS]

# Chapter 5: Solutions

**Section 5.1 Concept 1:** ____ *I can interpret probability as a long-run relative frequency.*

a) If you were to repeatedly draw cards, with replacement, from a shuffled deck, you would draw a jack, queen, or king 23% of the time.

b) No, we'd expect to draw a jack, queen, or king 23 times out of 100, but we are not guaranteed *exactly* 23.

**Section 5.1 Concept 2:** ____*I can use simulation to model chance behavior.*

Let the digits 01-95 represent a passenger showing up for the flight.
Let 96-00 represent a "no show".

Select 12 two digit numbers from a line of the random number table or a random number generator. Ignore repeats until you have 12 numbers selected. Count how many times 96-00 occurs. If 96-00 occurs less than two times, the flight is overbooked. Repeat this procedure 20 times. Determine how many of the 20 trials result in an overbooked flight.

**Section 5.1 Concept 1:** ____ *I can describe a probability model for a chance process*
____ *I can use basic probability rules such as the complement rule and addition rule for mutually exclusive events*

1. There are 52 possible outcomes. Each has a probability of 1/52.

2. P(A) = 4/52 = 0.0769. P(B) = 13/52 = 0.25.

3. $P(A^C)$ = 1-P(A) = 48/52 = 0.9231.

4. No, A and B are not mutually exclusive because a card can be both an Ace and a heart.

**Section 5.2 Concept 2:** ____ *I can use a Venn diagram to model a chance process involving two events* ____*I can find the probability of an event using a two-way table*
____ *I can use the general addition rule to calculate P(A U B)*

1. Use a two way table to display the sample space.

| A | | $A^C$ |
|---|---|---|
| B 1 | | 12 |
| $B^C$ 3 | | 36 |

2.

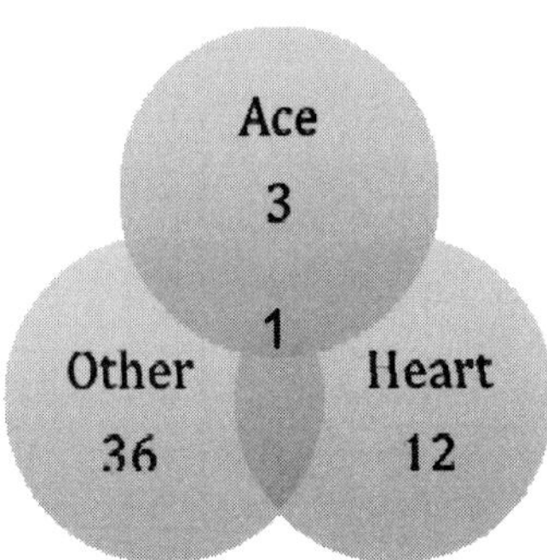


3. $P(A \cup B)$ = P(A) + P(B) – P(A and B) = 4/52 + 13/52 – 1/52 = 16/52 = 0.3077

**Section 5.3 Concept 1:** ____ *I can compute conditional probabilities*
____ *I can determine whether two events are independent*

If A and B are independent, P(A) = P(A|B). P(A) = 120/200 = 0.60. P(A|B) = 80/140 = 0.5714. Since these probabilities are not the same, A and B are not independent.

**Section 5.2 Concept 2:** ____ *I can use a tree diagram to describe chance behavior*
____ *I can use the general multiplication rule to solve probability questions*

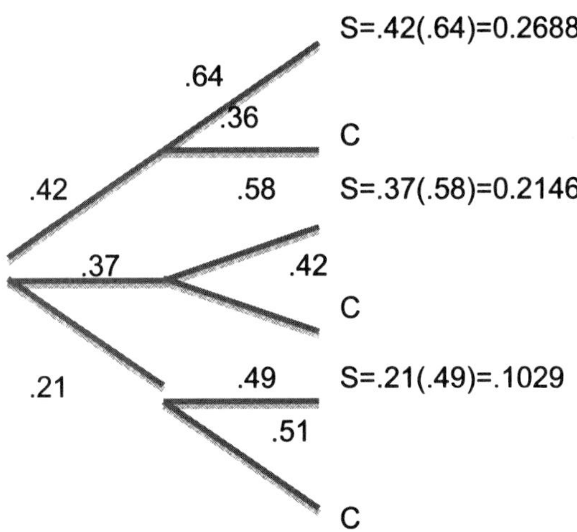


P(Statistics) = 0.2688+0.2146+0.1029 = 0.5863

**Section 5.3 Concept 3:** ____ *I can compute conditional probabilities*

P(Lakeville|Statistics) = 0.2688/0.5863 = 0.4585

# Chapter 5: Probability

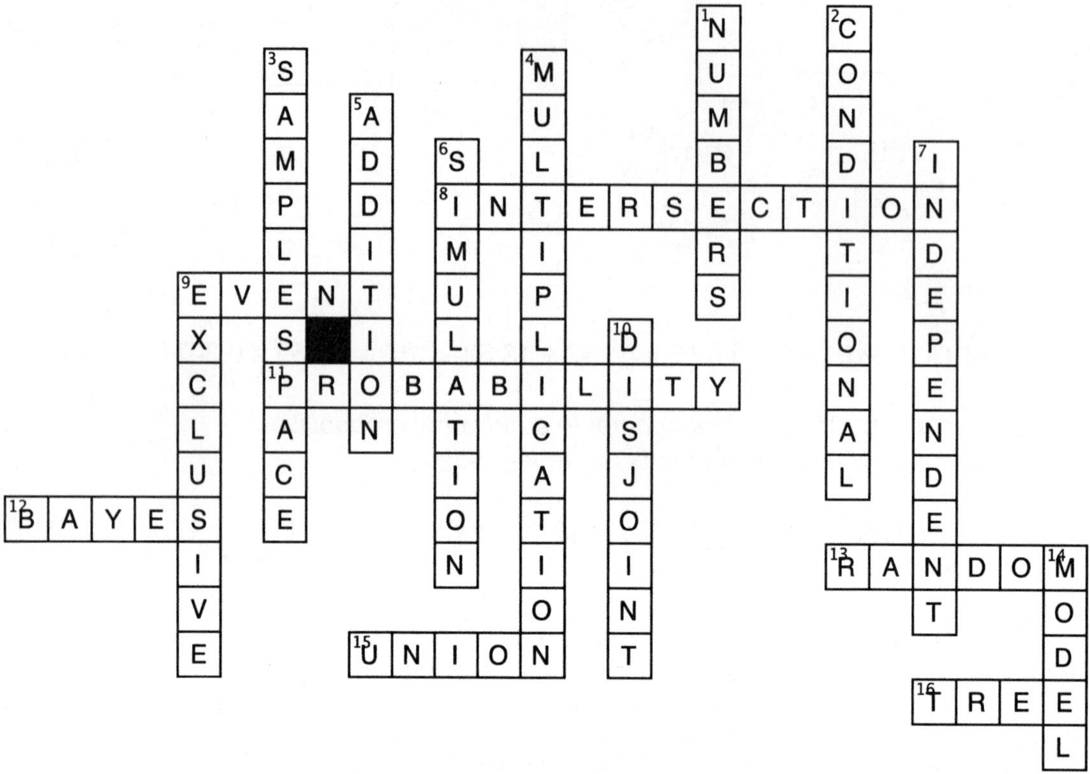

**Across**

8. The collection of outcomes that occur in both of two events. [INTERSECTION]
9. A collection of outcomes from a chance process. [EVENT]
11. The proportion of times an outcome would occur in a very long series of repetitions. [PROBABILITY]
12. _____ Theorem can be used to find probabilities that require going "backward" in a tree diagram. [BAYES]
13. In statistics, this doesn't mean "haphazard." It means "by chance." [RANDOM]
15. The collection of outcomes that occur in either of two events. [UNION]
16. A ____ diagram can help model chance behavior that involves a sequence of outcomes. [TREE]

**Down**

1. The law of large _____ states that the proportion of times an outcome occurs in many repetitions will approach a single value. [NUMBERS]
2. The probability that one event happens given another event is known to have happened. [CONDITIONAL]
3. The set of all possible outcomes for a chance process (two words). [SAMPLESPACE]
4. The probability that two events both occur can be found using the general _____ rule. [MULTIPLICATION]
5. P(A or B) can be found using the general _____ rule. [ADDITION]
6. The imitation of chance behavior, based on a model that reflects the situation. [SIMULATION]
7. The occurrence of one event has no effect on the chance that another event will happen. [INDEPENDENT]
9. Another term for disjoint: Mutually _____. [EXCLUSIVE]
10. Two events that have no outcomes in common and can never occur together. [DISJOINT]

# Chapter 6: Solutions

**Section 6.1 Concept 1:** ____ *I can use a probability distribution to answer questions about possible values of a random variable.*

Consider two 4-sided dice, each having sides labeled 1, 2, 3, 4. Let X = the sum of the numbers that appear after a roll of the dice.

a) X is a discrete random variable. We are most likely to roll a sum of 5 and least likely to roll a sum of 2 or 8.

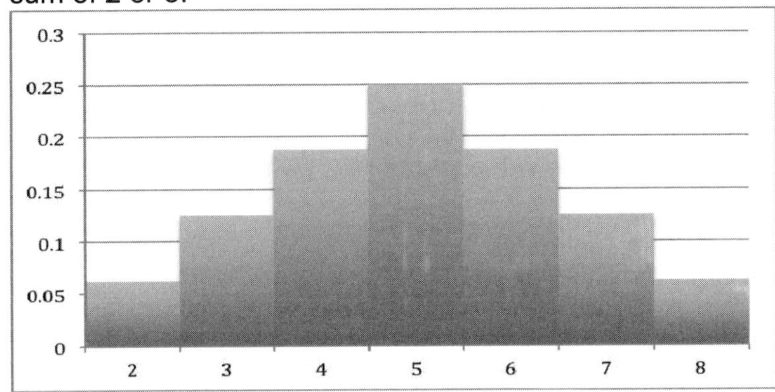


b) Yes, we should be surprised. In 10 rolls we would expect to see a sum of 3 or less about once or twice.

**Section 6.1 Concept 2:** ____ *I can calculate and interpret the mean of a random variable.*
____ *I can calculate and interpret the standard deviation of a random variable.*

E(Y) = 0(0.155)+ 1(0.195)+2(0.243)+3(0.233)+4(0.174) = 2.076. In the long run, we'd expect to see an average of 2.076 goals per game for many, many games.
VAR(Y) = $(0-2.076)^2(0.155)+\ldots+(4-2.076)^2(0.174) = 1.7382$
Standard deviation = $\sqrt{1.7382} = 1.3184$. We would expect the number of goals per game to vary by about 1.3184 from 2.076 in the long run.

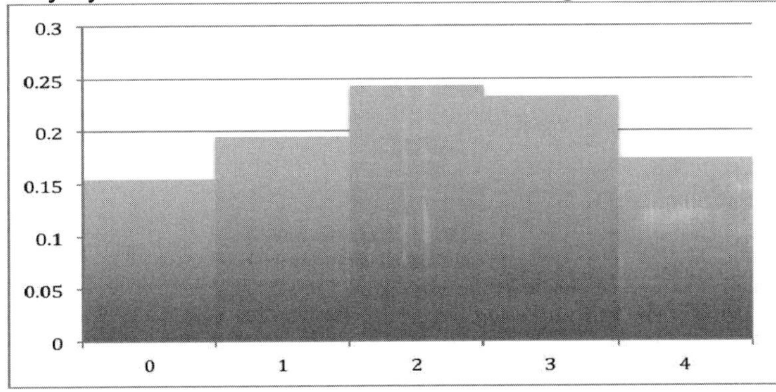


2. The weights of toddler boys follow an approximately Normal distribution with mean 34 pounds and standard deviation 3.5 pounds. Suppose you randomly choose one toddler boy and record his weight. What is the probability that the randomly selected boy weighs less than 31 pounds?

$z = (31-34)/3.5 = -0.8571$  $P(z<-0.8571) = 0.1956$.

**Section 6.2 Concept 1:** ____ *I can describe the effects of transforming a random variable.*

The shape will be slightly skewed to the right.
E(Y) = 1.5(E(X)) – 2 = -0.35. In the long run, we would expect to lose \$0.35 each time we play the game, on average.
StdDev(Y) = 1.5(0.943) = 1.4145. On average, we would expect our profit to vary by about \$1.42 around a loss of \$0.35.

**Section 6.2 Concept 2:** ____ *I can calculate and interpret the mean and standard deviation of the sum or difference of two random variables.*

1. Mean = 1.4+1.2+0.9+1 = 3.3 min
StdDev = $\sqrt{(0.1^2+0.4^2+0.8^2+0.7^2)} = \sqrt{1.3}$ = 1.14 min

2. Mean = 1 – 1.4 = -0.4 min (on average, Doug is faster by 0.4 min)
Std Dev = $\sqrt{(0.7^2+0.1^2)}$ = 0.7071 (The difference between Doug and Allan's times will vary by 0.7071 min around 0.4 min on average.)

**Section 6.2 Concept 3:** ____ *I can find probabilities involving the sum or difference of independent Normal random variables.*

Mean(M-L) = 110-100 = 10
StdDev(M-L) = $\sqrt{(10^2+8^2)}=\sqrt{164}$=12.81

Find P(M-L<0): z = (0-10)/12.81 = -0.78. P(z<-0.78) = 0.2177. There is about a 21.77% chance Mr. Molesky will finish before Mr. Liberty on any given day.

**Section 6.3 Concept 1:** ____ *I can determine whether the conditions for a binomial random variable have been met.* ____ *I can compute and interpret probabilities involving binomial distributions.*

B: A card is either a heart or it isn't
I: Each draw is independent since cards are replaced and the deck is shuffled
N: There are 10 observations in each game
S: The P(heart) = 0.25 in each draw.

P(X<4) = P(X=0)+P(X=1)+P(X=2)+P(X=3) = 0.7759

**Section 6.3 Concept 2:** ____ *I can calculate and interpret the mean and standard deviation of a binomial random variable.*

a) Show that X is approximately a binomial random variable.
B: Students either give a positive or negative rating
I: Since there are more than 10(500) students in the population, we can assume independence
N: 500 students are selected
S: P(positive rating) = 0.72 for each student

b) Use a Normal approximation to find the probability that 400 or more students would give their teacher a positive rating in this sample.

Mean = np = 500(.72) = 360
StdDev = $\sqrt{(np(1-p))}$ = 10.04
z = (400-360)/10.04 = 3.98    P(z ≥ 3.98) = 0.000034

**Section 6.3 Concept 3:** ____ *I can find probabilities involving geometric random variables.*

a) Show that X is a geometric random variable.
There are two outcomes (ring or no ring). Each box is independent. The probability of a ring in any given box is 0.2. We are interested in how long it will take to find a ring.

b) P(X=7) = $0.8^6(0.2)$ = 0.0524

c) P(X<4) = 0.488

d) E(X) = 1/.2 = 5 boxes.

## Chapter 6: Random Variables

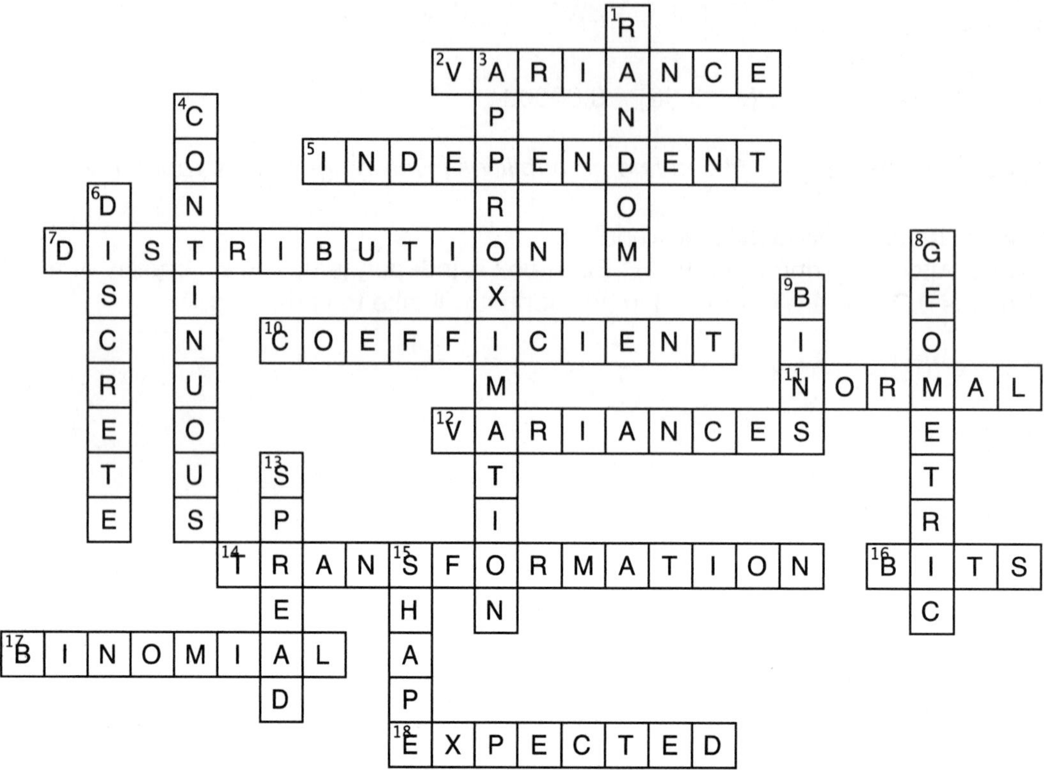


**Across**

2. The average of the squared deviations of the values of a variable from its mean. [VARIANCE]
5. Random variables are _____ if knowing whether an event in X has occurred tells us nothing about the occurrence of an event involving Y. [INDEPENDENT]
7. The probability _____ of a random variable gives its possible values and their probabilities. [DISTRIBUTION]
10. The number of ways of arranging k successes among n observations is the binomial ____. [COEFFICIENT]
11. The sum or difference of independent Normal random variables follows a _____ distribution. [NORMAL]
12. When you combine independent random variable, you always add these. [VARIANCES]
14. A linear _____ occurs when we add/subtract and multiply/divide by a constant. [TRANSFORMATION]
16. An easy way to remember the requirements for a geometric setting. [BITS]

**Down**

1. A ____ variable takes numerical values that describe the outcomes of some chance process. [RANDOM]
3. When n is large, we can use a Normal _____ to determine probabilities for binomial settings. [APPROXIMATION]
4. A random variable that takes on all values in an interval of numbers. [CONTINUOUS]
6. A random variable that takes a fixed set of possible values with gaps between. [DISCRETE]
8. A ____ setting arises when we perform independed trials of the same chance process and record the number of trials until a particular outcome occurs. [GEOMETRIC]
9. An easy way to remember the requirements for a binmial setting. [BINS]
13. Adding a constant to each value of a random variable has no effect on the shape or ____ of the distribution. [SPREAD]
15. Multiplying each value of a random variable by a constant has no effect on the ____ of the distribution. [SHAPE]

# Chapter 7: Solutions

**Section 7.1 Concept 1:** ____ *I can distinguish between a parameter and a statistic.*

a) Population: All males with high blood pressure
Parameter: Mean arterial pressure (μ) for all males with high blood pressure
Statistic: Sample mean arterial pressure for the 500 males in the study

b) Population: All 16- to 24-year old drivers
Parameter: Proportion of all 16- to 24-year old drivers who text while driving
Statistic: Sample proportion 0.12.

**Section 7.1 Concept 2:** ____ *I can distinguish between a population distribution, sampling distribution, and distribution of sample data.*

A breakfast cereal includes marshmallow shapes in the following distribution: 10% stars, 10% crescent moons, 20% rockets, 40% astronauts, 20% planets. We are interested in examining the proportion of rockets in a random sample of 2000 marshmallows from the cereal.

(a) The population distribution of marshmallow shapes.

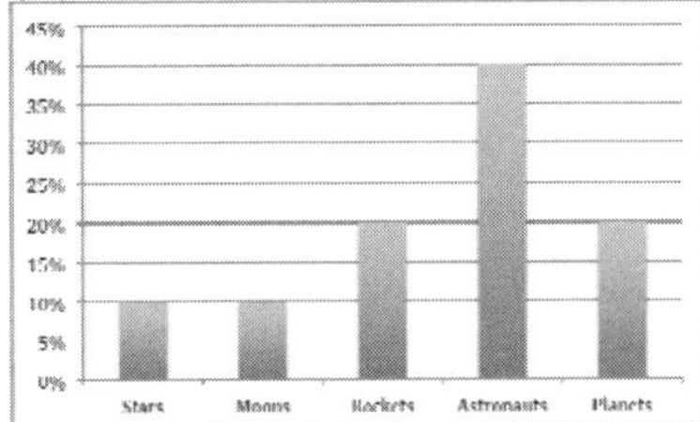


(b) The distribution of sample data we would expect to see (n=2000).

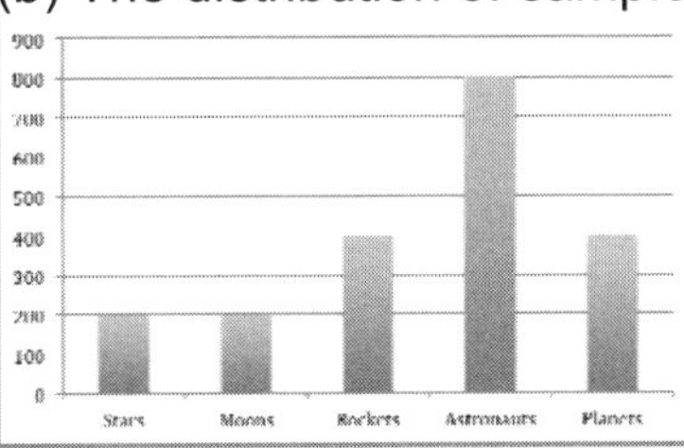


We would expect to see about 400 rockets.

(c) We would expect a symmetric distribution with mean 0.2 and standard deviation .009.

**Section 7.2 Concept 1:** ____ *I can calculate and interpret the mean and standard deviation of the sampling distribution of a sample proportion.* ____ *I can check whether the 10% and Normal conditions are met in a given setting.*

Suppose your job at a potato chip factory is to check each shipment of potatoes for quality assurance. Further, suppose that a truckload of potatoes contains 95% that are acceptable for processing. If more than 10% are found to be unacceptable in a random sample, you must reject the shipment. To check, you randomly select and test 250 potatoes. Let $\hat{p}$ be the sample proportion of unacceptable potatoes.

a) The mean of the sampling distribution of $\hat{p}$ is 0.05.

b) Both $np$=12.5 and $n(1-p)$=237.5 are greater than 10. The standard deviation of the sampling distribution of $\hat{p}$ is $\sigma_{\hat{p}} = \sqrt{\frac{.05(1-.05)}{250}} = .0137$.

c) Since both np and n(1-p) are greater than 10, we can assume the sampling distribution is approximately Normal. Since 0.10 is more than 3 standard deviations above the mean, it is unlikely we would observe a sample proportion greater than 10%. It is not likely we would reject the truckload based on a sample of 250 potatoes if 95% were truly acceptable.

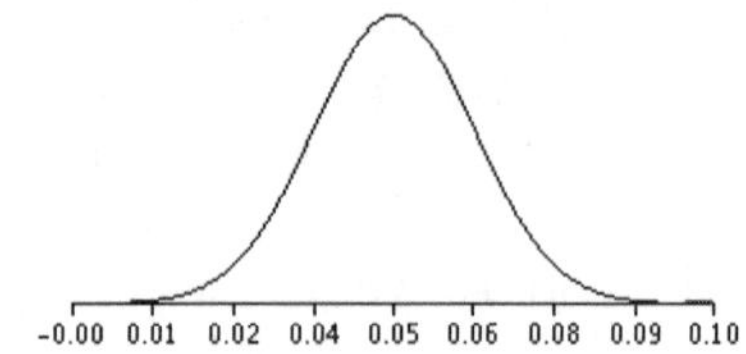


**Section 7.2 Concept 2:** ____ *I can use the Normal approximation to calculate probabilities involving sample proportions.* ____ *I can use the sampling distribution of $\hat{p}$ to evaluate a claim about a population proportion.*

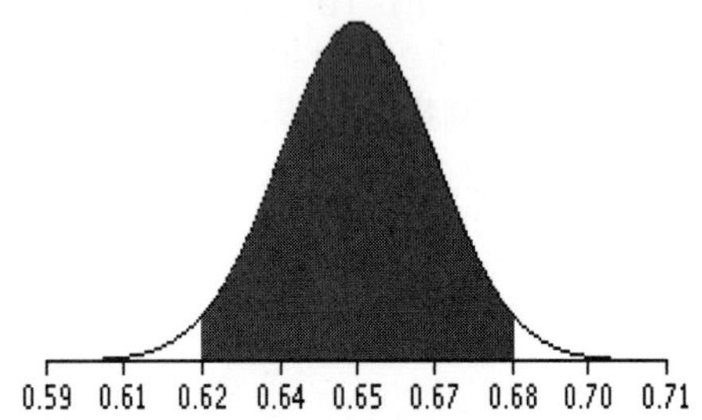


Since both np=650 and n(1-p)=350 are greater than 10, we can assume the sampling distribution of $\hat{p}$ is approximately Normal with mean 0.65 and standard deviation 0.015. 0.62 and 0.68 are 2 standard deviations above and below the mean. By the 68-95-99.7 rule, the probability a random sample of 1000 students will result in a $\hat{p}$ within 3-percentage points of the true proportion is approximately 95%.

**Section 7.3 Concept 1:** ____ *I can calculate and interpret the mean and standard deviation of the sampling distribution of a sample mean.* ____ *I can calculate probabilities involving a sample mean when the population distribution is Normal.*

P(a randomly chosen 5th grader will take more than 2.5 minutes

$z = \frac{2.5-2}{0.8} = 0.625$ P($z$>0.625)=0.2659

The sampling distribution of $\bar{x}$ will be N(2.5, 0.8/√20).

$z = \frac{2.5-2}{0.1789} = 2.795$ P($\bar{x}$>2.5)=0.002596

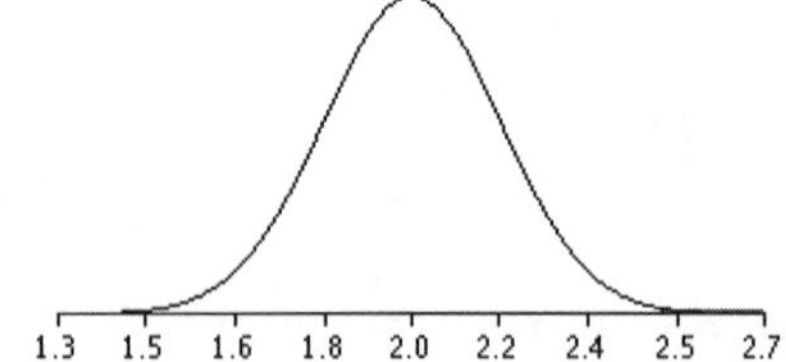


**Section 7.3 Concept 2:** ____ *I can use the central limit theorem to help find probabilities involving a sample mean.*

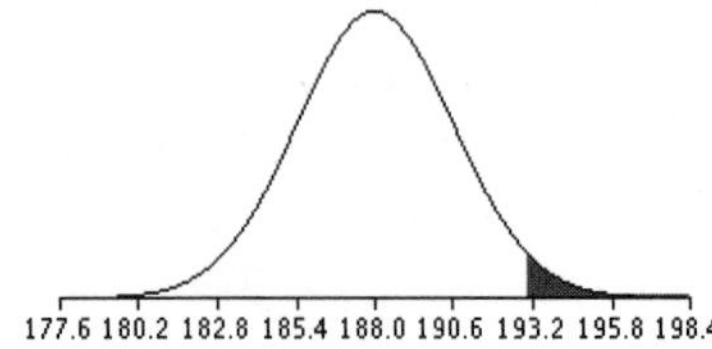

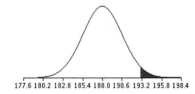

The sampling distribution of $\bar{x}$ will be N(188, 41/√250).

$z = \frac{193-188}{2.593} = 1.928$ P($\bar{x}$> 193)=0.0269.

# Chapter 7: Sampling Distributions

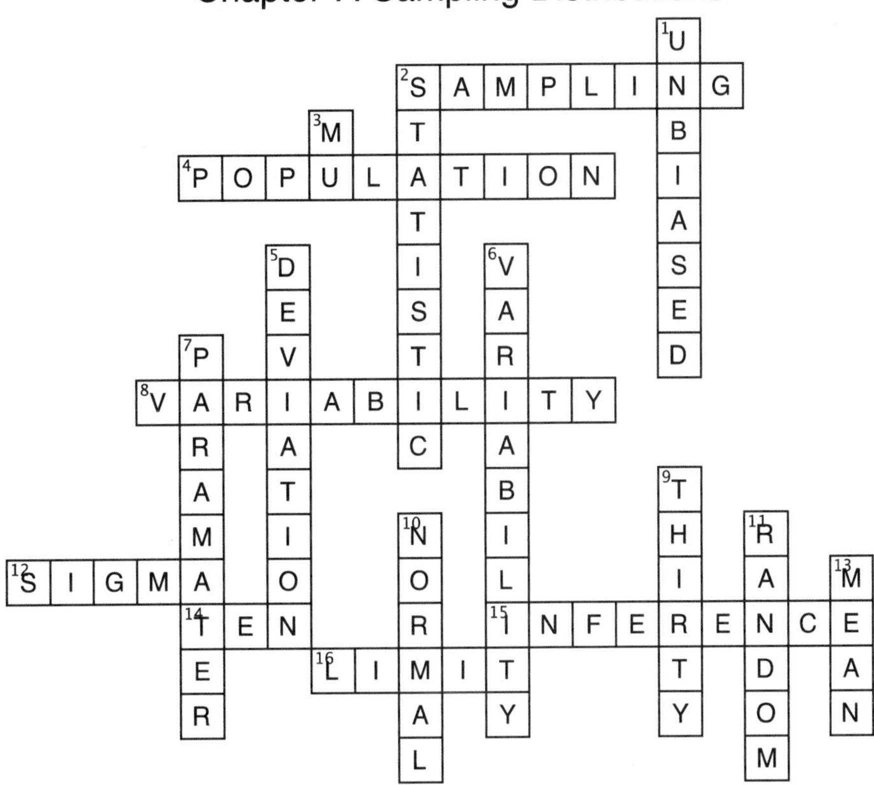

**Across**

2. _____ distribution: the distribution of values taken by the statistic in all possible samples of the same size from the population [SAMPLING]
4. _____ distribution: the distribution of all values of a variable in the population [POPULATION]
8. _____ of a statistic is described by the spread of the sampling distribution [VARIABILITY]
12. Greek letter used for the population standard deviation [SIGMA]
14. the Normal approximation for the sampling distribution of a sample proportion can be used when both the number of successes and failures are greater than _____ [TEN]
15. sampling distributions and sampling variability provide the foundation for performing _____ [INFERENCE]
16. central _____ theorem tells us if the sample size is large, the sampling distribution of the sample mean is approximately Normal, regardless of the shape of the population [LIMIT]

**Down**

1. a statistic is an _____ estimator if the mean of the sampling distribution is equal to the true value of the parameter being estimated. [UNBIASED]
2. a number, computed from sample data, that estimates a parameter [STATISTIC]
3. Greek letter used for the population mean [MU]
5. standard _____ : measure of spread of a sampling distribution [DEVIATION]
6. sampling _____ notes the value of a statistic may be different from sample to sample [VARIABILITY]
7. a number that describes a population [PARAMATER]
9. the rule of thumb for using the central limit theorem - the sample size should be greater than _____ [THIRTY]
10. when the sample size is large, the sampling distribution of a sample proportion is approximately _____ [NORMAL]
11. to draw a conclusion about a population parameter, we can look at information from a _____ sample [RANDOM]
13. center of a sampling distribution [MEAN]

# Chapter 8: Solutions

**Section 8.1 Concept 1:** ____ *I can interpret a confidence level.* ____ *I can interpret a confidence interval in context.* ____ *I can explain that a confidence interval gives a range of plausible values for the parameter.*

a) We are 90% confident the interval from 19.10 to 20.74 captures the true mean contents of a "20 oz." bottle of water.

b) If we were to collect many samples of 50 "20 oz." water bottles and construct a confidence interval for the mean contents in the same manner, 90% of the intervals would capture the true mean contents.

c) The average contents for the sample was 19.92 oz. Since 20 oz. is in the interval, we do not have evidence to suggest the population mean is less or greater than 20 oz.

**Section 8.1 Concept 2:** ____ *I can explain why each of the three inference conditions – random, Normal, and independent – is important.* ____ *I can explain how issues like nonresponse, undercoverage, and response bias can influence the interpretation of a confidence interval.*

Since there are more than 500 adults and since both np and n(1-p) are greater than 10, the Normal and Independent conditions are met. However, this may not be a random sample of the population of interest. While 50 addresses were randomly selected, the homes were called in late-morning. There is a possibility those who were home at that time are homemakers and, therefore, more likely to bake at least twice a week. The 90% confidence interval may overestimate the true proportion of adults who bake at least twice a week.

**Section 8.2 Concept 2:** ____ *I can construct and interpret a confidence interval for a population proportion.* ____ *I can determine critical values for calculating a confidence interval.*

State: We wish to construct a 90% confidence interval for the true proportion of adults who can roll their tongue.

Plan: We have a random sample, the number of successes and failures are both greater than 10 (68>10 and 232>10), and there are more than 3000 adults. We can construct a 90% confidence interval for the true proportion.

Do: $95\%CI = 0.23 \pm 1.645\sqrt{\frac{0.23(0.77)}{300}} = (0.18691, 0.26643)$

Conclude: We are 90% confident the interval from 0.19 to 0.27 captures the true proportion of adults who can roll their tongue.

**Section 8.2 Concept 3:** ____ *I can determine the sample size necessary to obtain a level C confidence interval for a population proportion with a specified margin of error.*

$$1.96 * \sqrt{\frac{.5(.5)}{n}} \le 0.02 \Rightarrow \sqrt{\frac{.5(.5)}{n}} \le 0.0102 \Rightarrow \frac{0.25}{n} \le 0.000104$$

$$n \ge 2401$$

**Section 8.3 Concept 2:**
Use the *t* table to determine the critical value *t** that you would use for a confidence interval.
a) $t^*=1.330$ b) $t^*=1.984$ c) $t^*=2.756$

**Section 8.3 Concept 3:** ____ *I can construct and interpret a confidence interval for a population mean.*

This is a random sample, however the sample size is less than 30. A boxplot of the sample data does not suggest strong skewness or outliers, so we can construct a 95% confidence interval for the true mean.

0 5 10 15 20 25 30 35
sugar

$$95\%CI : 22.5 \pm 2.365\frac{7.191}{\sqrt{8}} = (16.488, 28.512)$$

We are 95% confident the interval from 16.488 to 28.512 captures the true mean amount of sugar for this manufacturer's soft drinks.

$$z^* \frac{\sigma}{\sqrt{n}} \leq ME$$

**Section 8.3 Concept 4:** ____ *I can determine the sample size required to obtain a level C confidence interval for a population mean with a specified margin of error.*

$$1.96 * \frac{4}{\sqrt{n}} \leq 0.5 \Rightarrow \frac{4}{\sqrt{n}} \leq 0.2551 \Rightarrow 15.68 \leq \sqrt{n}$$

$$n \geq 245.86$$

# Chapter 8: Estimating with Confidence

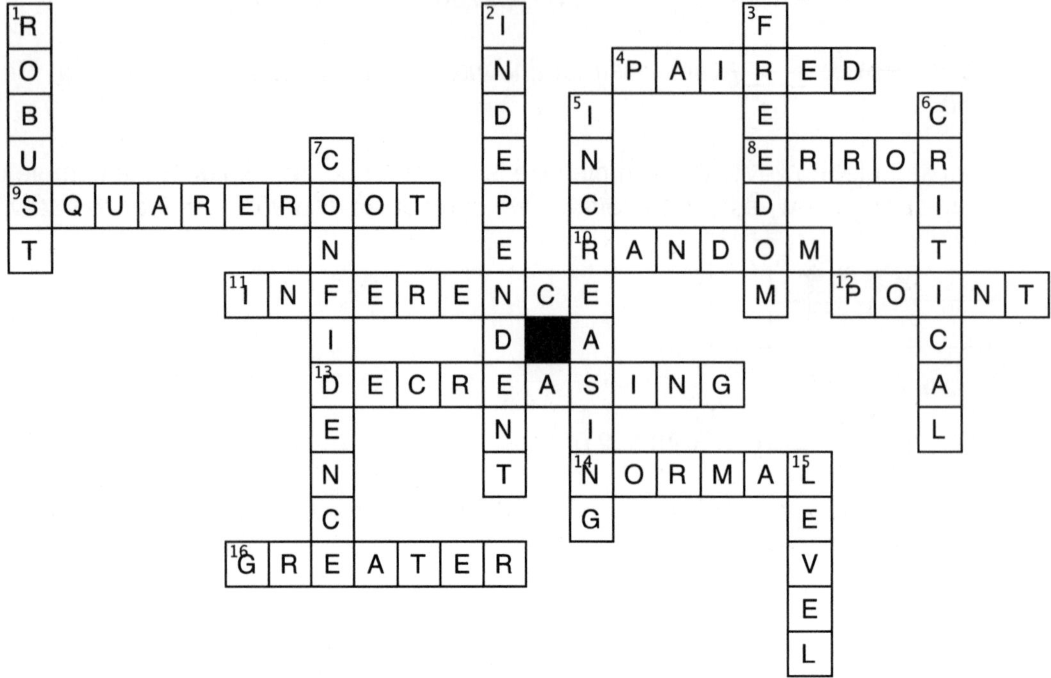

**Across**

4. ____ t procedures allow us to compare the responses to two treatments in a matched pairs design [PAIRED]
8. a confidence interval consists of an estimate ± margin of ____ [ERROR]
9. to find the standard error of the sample mean, divide the sample standard deviation by the ____ of the sample size (two-words) [SQUAREROOT]
10. to estimate with confidence, our estimate should be calculated from a ___ sample [RANDOM]
11. methods for drawing conclusions about a population from sample data [INFERENCE]
12. a single value used to estimate a parameter is a ____ estimator [POINT]
13. we can construct a narrow interval by ____ our confidence [DECREASING]
14. as degrees of freedom increase, the t distribution approaches the ____ distribution [NORMAL]
16. the spread of the t distributions is ____ than the spread of the standard Normal distribution [GREATER]

**Down**

1. inference procedures that remain fairly accurate even when a condition is violated [ROBUST]
2. another condition for confidence intervals is that observations should be ____ [INDEPENDENT]
3. particular t distributions are specified by degrees of ____ [FREEDOM]
5. we can construct a narrow confidence interval by ____ our sample size [INCREASING]
6. the margin of error consists of a ____ value and the standard error of the sampling distribution [CRITICAL]
7. a ____ interval provides an estimate for a population parameter [CONFIDENCE]
15. confidence ____: the success rate of the method in repeated sampling [LEVEL]

# Chapter 9: Solutions

**Section 9.1 Concept 1:** ____ *I can state correct hypotheses for a significance test about a population proportion or mean.*

a) The parameter of interest is the true proportion of hearts in the "chute" of cards.
b) $H_0$: $p = 0.25$ vs. $H_a$: $p > 0.25$
c) The sample proportion is 7/12 = .0583. It is possible to be dealt 7 hearts in 12 cards, however it is very unlikely. We would expect about 3 hearts when dealt 12 cards.

**Section 9.1 Concept 2:** ____ *I can interpret P-values in context.*

a) If the null hypothesis was true, the proportion of hearts in the "chute" would be 0.25.
b) Assuming the proportion of hearts in the "chute" is 0.25, there is a 0.4% chance we would observe 7 or more hearts when 12 cards were randomly selected. This is highly unlikely.
c) Since it is so unlikely that we would observe 7 or more hearts when dealt 12 cards from a fair chute (less than a 5% chance), we have evidence to suggest there may be more hearts than usual in the chute.

**Section 9.1 Concept 3:** ____ *I can interpret a Type I error and a Type II error in context, and give the consequences of each.* ____ *I can describe the relationship between the significance level of a test, P(Type II error), and power.*

a) A Type I error would occur if we assumed the chute had more hearts than usual (based on our sample) when, in reality, the proportion of hearts was 0.25. We were just dealt an unusual hand.
b) A Type II error would occur if we assumed the chute was fair when, in reality, the proportion of hearts was actually greater than 0.25.

**Section 9.2 Concept 1:** ____ *I can check conditions for carrying out a test about a population proportion.* ____ *I can conduct a significance test about a population proportion.*

$p$ = the true proportion of passages that follow the speech pattern in the work in question
$H_0$: $p = 0.214$
$H_a$: $p > 0.214$

Conditions: We have a random sample of passages from the work in question. The number of successes and failures (136 and 303) are both greater than 10.

Assuming the null hypothesis is true, the sampling distribution of the proportion of passages following the speech pattern will be N(0.214, 0.0196).

$z = \dfrac{0.3098 - 0.214}{0.0196} = 4.89$   P-value = 0.000001

Since the P-value is less than 5%, we have significant evidence to reject the null hypothesis. It appears the work in question may have too high a proportion of passages following Plato's speech pattern to be one of his actual works.

**Section 9.2 Concept 2:** ____ *I can use a confidence interval to draw a conclusion for a two-sided test about a population proportion.*

We wish to construct a 99% confidence interval for the true proportion of teens who text while driving. We have a random sample, there are more than 270 teenage drivers and the number of successes and failures (15 and 12) are both greater than 10.

$$99\%CI : 0.56 \pm 2.576\sqrt{\frac{0.56(0.44)}{27}} = (0.30923, 0.80188)$$

We are 99% confident the interval from 0.31 to 0.80 captures the true proportion of teens who text while driving. Since 0.77 is in this interval, we do not have evidence to suggest the true proportion is different than 77%.

**Section 9.3 Concept 1:** ____ *I can check conditions for carrying out a test about a population mean.* ____ *I can conduct a one-sample t test about a population mean μ.*

We will conduct a one-sample *t*-test for the true mean ratio.
$H_0$: $\mu = 8.9$
$H_a$: $\mu \neq 8.9$

Conditions: We have a random sample of 41 bones and *n*>30.

Assuming the null hypothesis is true, the sampling distribution will be *t* with 40 degrees of freedom. The mean of the sampling distribution is 8.9 and the standard error is 0.187.

$$t = \frac{9.27 - 8.9}{1.198/\sqrt{41}} = 1.97$$ P-value = 0.056

Since the P-value is greater than 5%, we do not have significant evidence to reject the null hypothesis. There is no evidence at the 5% level to suggest these bones have a ratio different than 8.9.

**Section 9.3 Concept 2:** ____ *I can construct and interpret a confidence interval for a population mean.*

The mean difference is 20.18. We will conduct a matched pairs *t*-test for the true mean difference between the caffeine and placebo beats per minute.
$H_0$: $\mu_d = 0$
$H_a$: $\mu_d > 0$

Conditions: The treatments were randomly assigned. We do not have more than 30 subjects. A boxplot of the differences suggests several outliers so we will proceed with caution.

$$t = \frac{20.18 - 0}{48.75/\sqrt{11}} = 1.37$$ P-value = 0.099.

Since the P-value is greater than 5%, we do not have sufficient evidence to reject the null hypothesis. We cannot conclude that caffeine results in a higher beats per minute than the placebo.

## Chapter 9: Testing a Claim

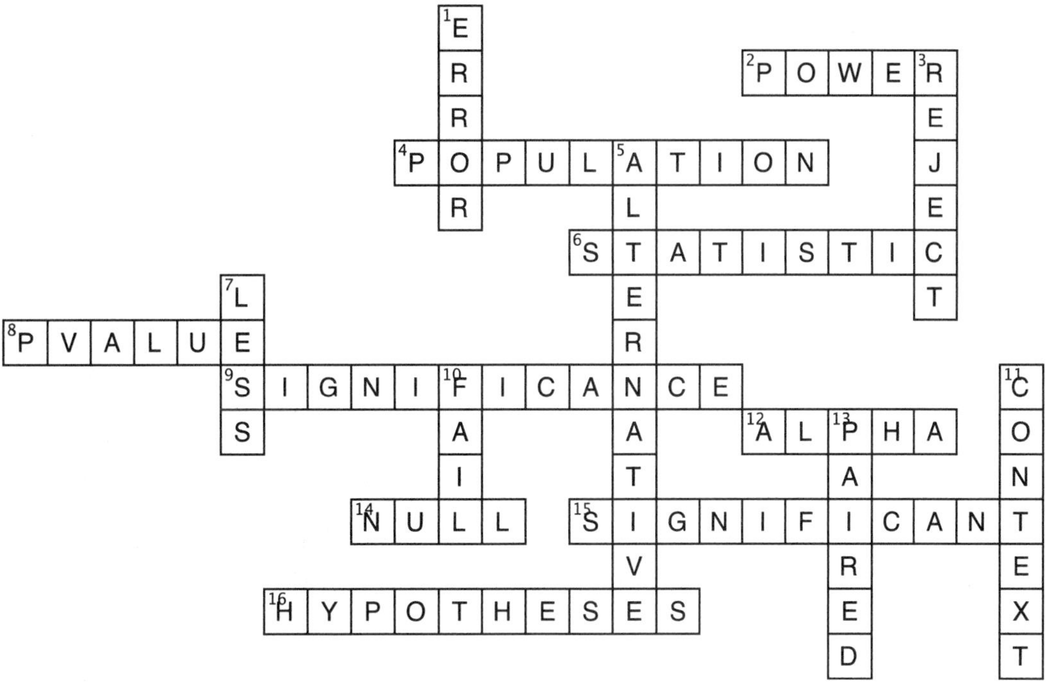


**Across**

2. the probability that a significance test will reject the null when a particular alternative value of the paramter is true [POWER]
4. hypotheses always refer to the ____ [POPULATION]
6. the test ____ is a standardized value that assesses how far the estimate is from the hypothesized parameter [STATISTIC]
8. the probability that we would observe a statistic at least as extreme as the one observed, assuming the null is true (two terms) [PVALUE]
9. we can use a _____ test to compare observed data with a hypothesis about a population [SIGNIFICANCE]
12. greek letter used to designate the significance level [ALPHA]
14. the ____ hypothesis is a the claim for which we are seeking evidence against [NULL]
15. an observed difference that is too small to have occured due to chance alone is considered statistically ____ [SIGNIFICANT]
16. the statements a statistical test is designed to compare [HYPOTHESES]

**Down**

1. if we reject the null hypothesis when it is actually true, we commit a Type I ____ [ERROR]
3. if we calculate a very small P value, we have evidence to ____ the null [REJECT]
5. the ____ hypothesis is the claim about the population for which we are finding evidence for [ALTERNATIVE]
7. reject the null hypothesis if the P value is ____ than the significance level [LESS]
10. if our calculated P value is not small enough to provide convincing evidence, we ___ to reject he null [FAIL]
11. conclusions should always be written in ____ [CONTEXT]
13. a ____ test allows us to analyze differences in responses within pairs [PAIRED]

# Chapter 10: Solutions

**Section 10.1 Concept 1:** ____ *I can describe the characteristics of the sampling distribution of $\hat{p}_1 - \hat{p}_2$* ____ *I can calculate probabilities using the sampling distribution of $\hat{p}_1 - \hat{p}_2$*

a) The sampling distribution of $\hat{p}_S - \hat{p}_N$ will be approximately Normal with a mean of 0.15 and a standard deviation of 0.0556.

b) $z = \dfrac{.07 - .15}{\sqrt{\dfrac{.25(.75)}{160} + \dfrac{.6(.4)}{125}}} = -1.439$ P(z<-1.439) = 0.0749.

c) Assuming the stated proportions for each high school are true, there is a 7.5% chance we'd observe differences between the sample proportions at least as extreme as those observed. We do not have evidence to doubt the study's sample proportions.

**Section 10.1 Concept 2:** ____ *I can determine whether or not the conditions for performing inference are met.* ____ *I can construct and interpret a confidence interval to compare two proportions.*

Since we have two random samples, more than 10 successes in each sample (551 in 1990 and 652 in 2010), more than 10 failures in each sample (949 in 1990 and 1348 in 2010), and we sampled less than 10% of the population of interest each year, we can construct a 95% confidence interval for the difference in proportions.

$$95\% CI: (0.367 - 0.326) \pm 1.96\sqrt{\frac{0.367(1-0.367)}{1500} + \frac{0.326(1-0.326)}{2000}} = (0.0094, 0.0732)$$

We are 95% confident the interval from 0.0094 t0 0.0732 captures the true difference in the proportions of adults who smoked in 1990 and in 2010. Since 0 is not contained in this interval, we can conclude the proportion of adults who smoked in 1990 was higher than in 2010.

**Section 10.1 Concept 3:** ____ *I can perform a significance test to compare two proportions.*

We will perform a two-sample test for proportions.
$H_0$: $p_N = p_S$
$H_a$: $p_N > p_S$

Conditions: We have random sample from each school. The number of successes and failures at North (28 and 92) are both greater than 10. The number of successes and failures at South (30 and 120) are both greater than 10. There are at least 1200 students at North and at least 1500 students at South.

$z = \dfrac{0.233 - 0.2}{\sqrt{\dfrac{.2148(1-.2148)}{120} + \dfrac{.2148(1-.2148)}{150}}} = 0.6626$ P-Value=0.2537

Since the P-value is greater than 5%, we do not have evidence to reject the null hypothesis. We cannot conclude that the proportion of low income students at North is higher than the proportion of low income students at South.

**Section 10.2 Concept 1:** ____ *I can describe the characteristics of the sampling distribution of $\bar{x}_1 - \bar{x}_2$* ____ *I can calculate probabilities using the sampling distribution of $\bar{x}_1 - \bar{x}_2$*

a) The sampling distribution of $\bar{x}_1 - \bar{x}_2$ will be a *t* distribution (df = 58.79). The mean will be 24 and the standard deviation will be $\sqrt{\frac{18^2}{40}+\frac{9.4^2}{40}} = 3.21$.

b)

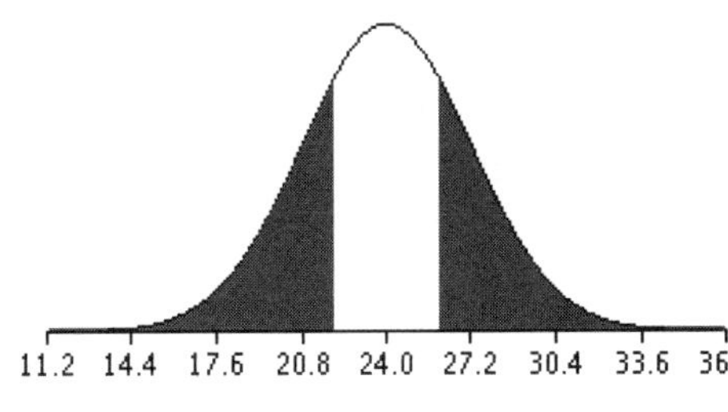


$$t = \frac{22-24}{3.21} = -0.62$$
$$t = \frac{26-24}{3.21} = 0.62$$

P(difference in means is 2 points or more) = 2(0.269)=0.538 (approximate, based on df=58.79).

**Section 10.2 Concept 2:** ____ *I can construct and interpret a two-sample t interval to compare two means.*

We will construct a 95% confidence interval for the difference in mean weight loss.
Conditions: Patients were randomly assigned to each treatment. We have at least 30 patients on each treatment.

$$95\%CI: (9.3-7.4) \pm t^* \sqrt{\frac{4.7^2}{100}+\frac{4^2}{100}} = (0.6827, 3.117)$$

We are 95% confident the interval from 0.68 to 3.12 captures the true difference in mean weight loss for the two diets. Since 0 is contained in this interval, we do not have evidence to suggest the new diet is more effective than the current diet.

**Section 10.2 Concept 3:** ____ *I can perform a two-sample t test to compare two means.*

We will perform a two-sample *t* test for the difference in mean memory scores.
$H_0$: $\mu_B = \mu_G$
$H_a$: $\mu_B > \mu_G$

Conditions: We have two independent random samples of boys and girls. Both samples have at least 30 individuals.

$$t = \frac{48.9-48.4}{\sqrt{\frac{12.96}{200}+\frac{11.85}{150}}} = 0.375$$ P-value = 0.354

Since the P-value is greater than 5%, we do not have sufficient evidence to reject the null hypothesis. We cannot conclude that the boys have better short term memory than the girls.

# Chapter 10: Comparing Two Populations or Groups

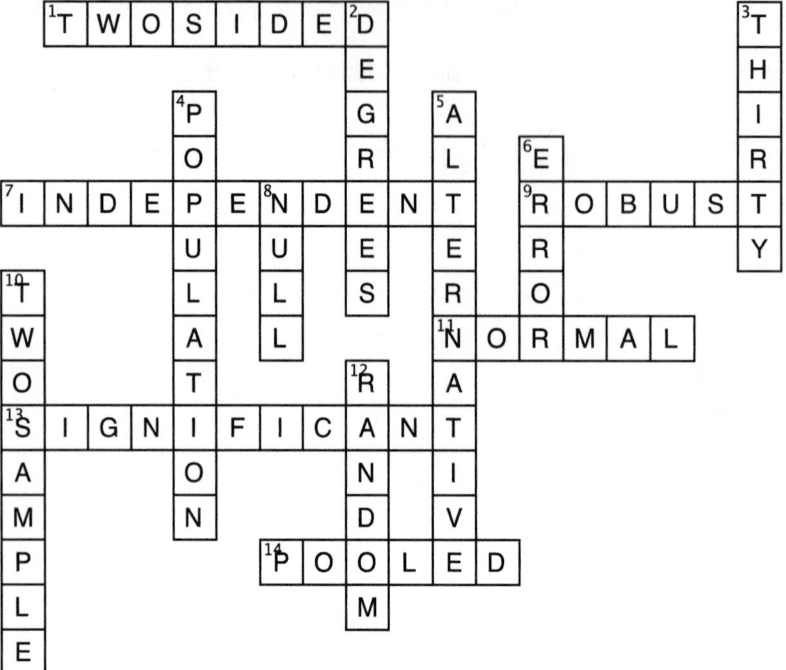

**Across**

1. an alternative hypothesis the seeks evidence of a difference requires the use of a ____ test (two words) [TWOSIDED]
7. to perform two-sample procedures, the random samples should be____ [INDEPENDENT]
9. procedures that yield accurate results, even when a condition is violated, are ____ [ROBUST]
11. if the number of successes and failures from both samples are greater than ten, the sampling distribution for the difference between the two proportions will be approximately ____ [NORMAL]
13. results that are too unlikely to be due to chance alone are considered statistically ____ [SIGNIFICANT]
14. in a two-proportion test, we calculate a ____ proportion [POOLED]

**Down**

2. particular t distributions are distinguished by ____ of freedom [DEGREES]
3. the sampling distribution for the difference between means will be approximately Normal if both sample sizes are greater than [THIRTY]
4. hypotheses should always be written in terms of the ____ [POPULATION]
5. the hypthesis we are seeking evidence for [ALTERNATIVE]
6. the margin of error in a confidence interval consists of a critical value and standard ____ of the statistic [ERROR]
8. the hypothesis we are seeking evidence against [NULL]
10. to compare two proportions, we can construct a ____ z interval (two words) [TWOSAMPLE]
12. if we want to compare two population proprtions, we must have data from two ____ samples [RANDOM]

# Chapter 11: Solutions

**Section 11.1 Concept 1:** ____*I can compute expected counts and contributions to the chi-square statistic.*

$$\chi^2 = \frac{(42-50)^2}{50} + \frac{(55-50)^2}{50} + \frac{(38-50)^2}{50} + \ldots + \frac{(44-50)^2}{50}$$

$$\chi^2 = 10.28$$

**Section 11.1 Concept 2:** ____ *I can check the Random, Large Sample Size, and Independent conditions before performing a chi-square test.* ____ *I can use a chi-square goodness-of-fit test to determine whether sample data are inconsistent with a specified distribution of a categorical variable.* ____ *I can examine individual components of the chi-square statistic as part of a follow-up analysis.*

We will perform a chi-square goodness-of-fit test.
$H_0$: Each value has an equal probability (1/6) of occurring
$H_a$: At least one value does not have a 1/6 chance of occurring

We have a random sample since the die was rolled 300 times. All expected counts (50) are greater than 5 and each roll is independent.

Chi-square = 10.28. With df=5, the P-value will be between .05 and 0.10.

Since the P-value is greater than 5%, we do not have sufficient evidence to reject the null hypothesis. We cannot conclude that the die is not fair.

**Section 11.2 Concept 2:** ____ *I can use a chi-square test for homogeneity to determine whether the distribution of a categorical variable differs for several populations or treatments.*

| | Zooboomafoo | iCarly | Phineas and Ferb |
|---|---|---|---|
| Boys | 20 (30) | 30 (36) | 50 (33.3) |
| Girls | 70 (60) | 80 (73.3) | 50 (66.6) |

Do these data provide convincing evidence that television preferences differ significantly for boys and girls?

We will perform a chi-square test for homogeneity.
$H_0$: the distribution of preferences is the same for boys and girls
$H_a$: the distributions of preference are not the same

Conditions: We have two independent random samples. All expected counts are greater than 5.

Chi-square = 19.32. With df=2, the P-value = $6.38\text{x}10^{-5}$

Since the P-value is less than 5%, we have significant evidence to reject the null hypothesis. It appears the distribution of preferences for boys and girls are not the same. A follow-up analysis suggests the biggest difference occurs in the preferences for Phineas and Ferb, with more boys and fewer girls preferring that show than expected.

**Section 11.2 Concept 3:** ____ *I can use a chi-square test of association/independence to determine whether there is convincing evidence of an association between two categorical variables.*

A recent study looked into the relationship between political views and opinions about nuclear energy. A survey administered to 100 randomly selected adults asked their political leanings as well as their approval of nuclear energy. The results are below:

| | Liberal | Conservative | Independent |
|---|---|---|---|
| Approve | 10 (12.15) | 15 (8.55) | 20 (24.3) |
| Disapprove | 9 (7.29) | 2 (5.13) | 16 (14.58) |
| No Opinion | 8 (7.56) | 2 (5.32) | 18 (15.12) |

We will perform a chi-square test for association/independence.
$H_0$: there is no association between political leanings and views on nuclear energy
$H_a$: there is an association between political leanings and views on nuclear energy

Conditions: We have a random sample of adults. All expected counts are greater than 5 and there are at least 1000 adults in the population.

Chi-square = 11.10. With df=4, the P-value = 0.025.

Since the P-value is less than 5%, we have significant evidence to reject the null hypothesis. It appears there is an association between political leanings and views on nuclear energy. A follow-up analysis suggests the biggest difference occurs with more Conservatives approving of nuclear energy than expected.

## Chapter 11: Inference for Distributions of Categorical Data

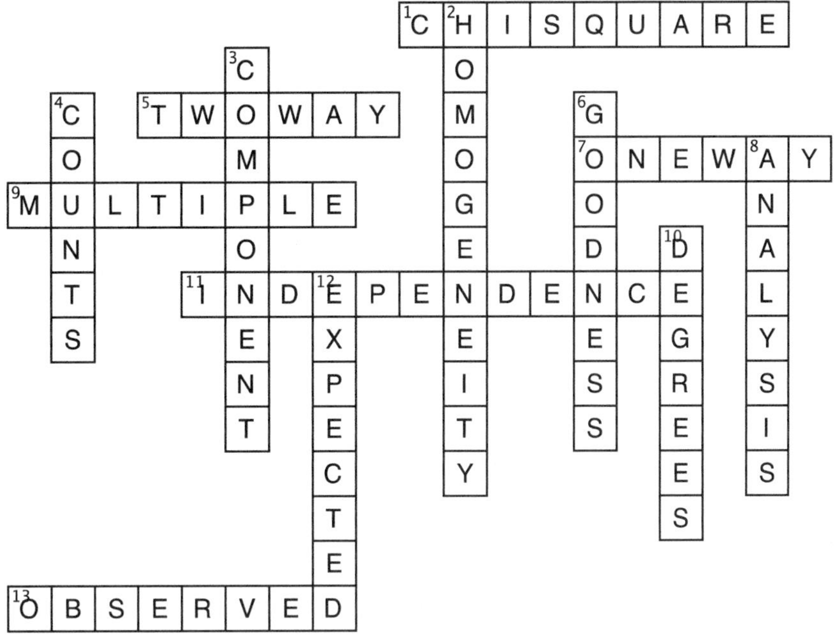


### Across

1. test statistic used to test hypotheses about distributions of categorical data (two words) [CHISQUARE]
5. when comparing two or more categorical variables, or one categorical variable over multiple groups, we arrange our data in a ______ table (two words) [TWOWAY]
7. a ____ table summarizes the distribution of a single categorical variable (two words) [ONEWAY]
9. the problem of doing many comparisons at once with an overall measure of confidence is the problem of ____ comparisons [MULTIPLE]
11. to test whether there is an association between two categorical variables, use a chi-square test of association / ____ [INDEPENDENCE]
13. type of categorical count gathered from a sample of data [OBSERVED]

### Down

2. to test whether there is a difference in the distribution of a categorical variable over several populations or treatments, use a chi-square test of ____ [HOMOGENEITY]
3. a follow up analysis involves identifying the ____ that contribued the most to the chi-square statistic [COMPONENT]
4. when calculating chi-square statistics, observations should be expressed in ____, not percents [COUNTS]
6. ____ of fit test: used to determine whether a population has a certain hypothesized distribution of proportions for a categorical variable [GOODNESS]
8. after conducting a chi-square test, be sure to carry out a follow-up ____ [ANALYSIS]
10. specific chi-square distributions are distinguished by ____ of freedom [DEGREES]
12. type of categorical count that we would see if the null hypothesis were true [EXPECTED]

# Chapter 12: Solutions

**Section 12.1 Concept 1:** ____ *I can construct and interpret a confidence interval for the slope of a least-squares regression line.*

$95\%CI: -4.77 \pm 2.306(1.0105) = (-7.10, -2.44)$
We are 95% confident the interval from -7.10 to -2.44 captures the true slope of the regression line.

**Section 12.1 Concept 2:** ____ *I can perform a significance test about the slope of a population (true) regression line.* ____ *I can interpret computer output from a least-squares regression analysis.*

The study in the previous Check for Understanding was expanded to include a total of 77 randomly selected cereals. The scatterplot, residual plot, and computer output of the regression analysis are noted above. Use this output to determine the LSRL for the sample data.

LSRL: predicted rating score = 59.284 – 2.4008(sugar)

The slope = -2.400. For each increase of 1g of sugar, we estimate approximately a 2.4 point decrease in rating score.

Is there convincing evidence that the slope of the true regression line is less than zero?
We will perform a *t*-test for the slope of the regression line.
$H_0: \beta = 0$
$H_a: \beta < 0$

Conditions: The cereals are randomly selected and independent. The scatterplot suggests a negative linear relationship. The residual plot does not display a distinct pattern, although the residuals do appear to be smaller for greater sugar content. We will proceed with caution.

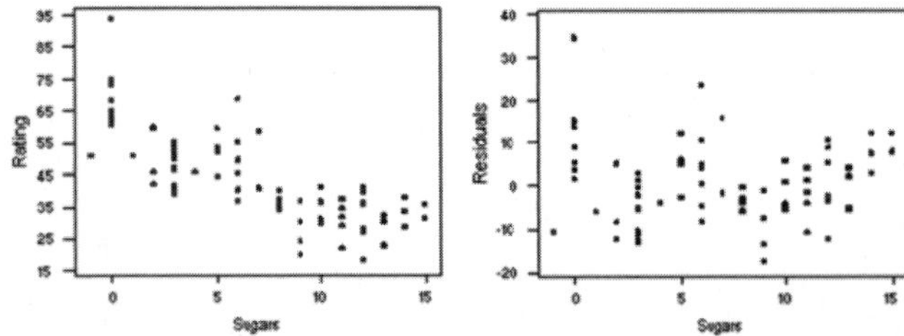


According to the computer output, $t$=-10.12 and the P-Value = 0.000.

Since the P-value is less than 5%, we have significant evidence to reject the null hypothesis. It appears the slope of the true regression line may be less than zero.

**Section 12.2 Concept 1:** ____ *I can use transformations involving powers and roots to achieve linearity for a relationship between two quantitative variables.*

Plot (diameter$^{0.7}$, length)

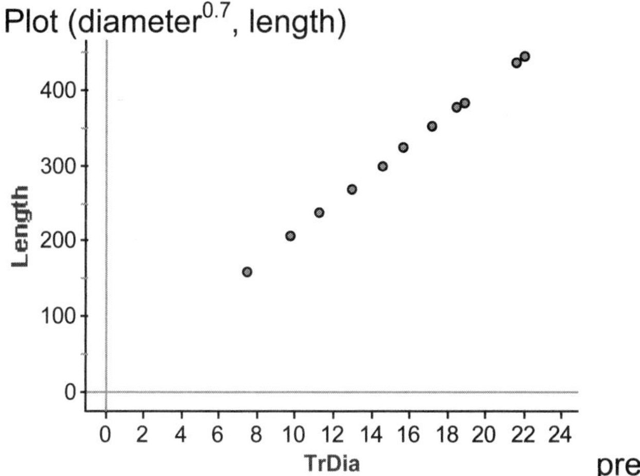


predicted length = 19.5(diameter)$^{0.7}$ +17

If diameter = 47mm, predicted length = 19.5(47)$^{0.7}$ +17 = 305.74.

**Section 12.2 Concept 2:** ____ *I can use transformations involving logarithms to achieve linearity for a relationship between two quantitative variables.* ____ *I can determine which of several transformations does a better job of producing a linear relationship.* ____ *I can make predictions from a least-squares regression line involving transformed data.*

The following data describe the number of police officers (thousands) and the violent crime rate (per 100,000 population) in a sample of states. Use these data to determine a model for predicting violent crime rate based on number of police officers employed. Show all appropriate plots and work.

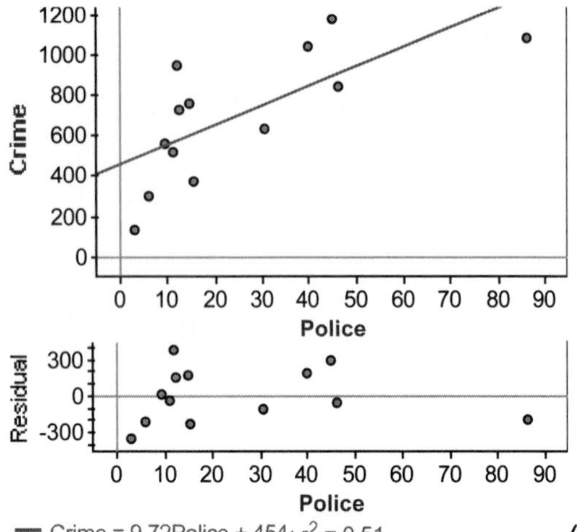


— Crime = 9.72Police + 454; $r^2 = 0.51$

(police, crime) displays a slightly curved pattern.

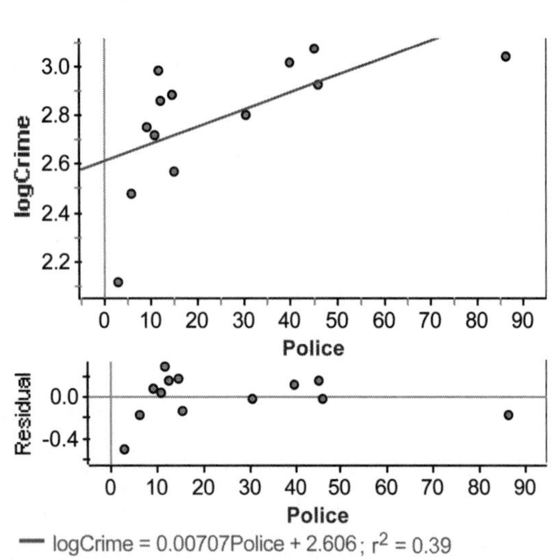


— logCrime = 0.00707Police + 2.606; $r^2 = 0.39$

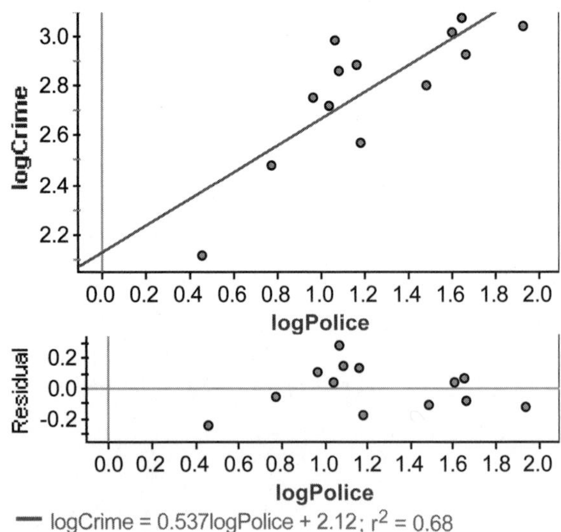


— logCrime = 0.537logPolice + 2.12; $r^2 = 0.68$

(log(police), log(crime)) is the most linear of the plots. Therefore, a power model may be the best choice to describe the relationship between police and crime.

(log(crime)-hat) = 2.12 + 0.537(log(police))

If a state has 25,400 police officers,
log(crime) = 2.12 + 0.537(log(25.4))
log(crime) = 2.874
crime = 748.85

# Chapter 12: More About Regression

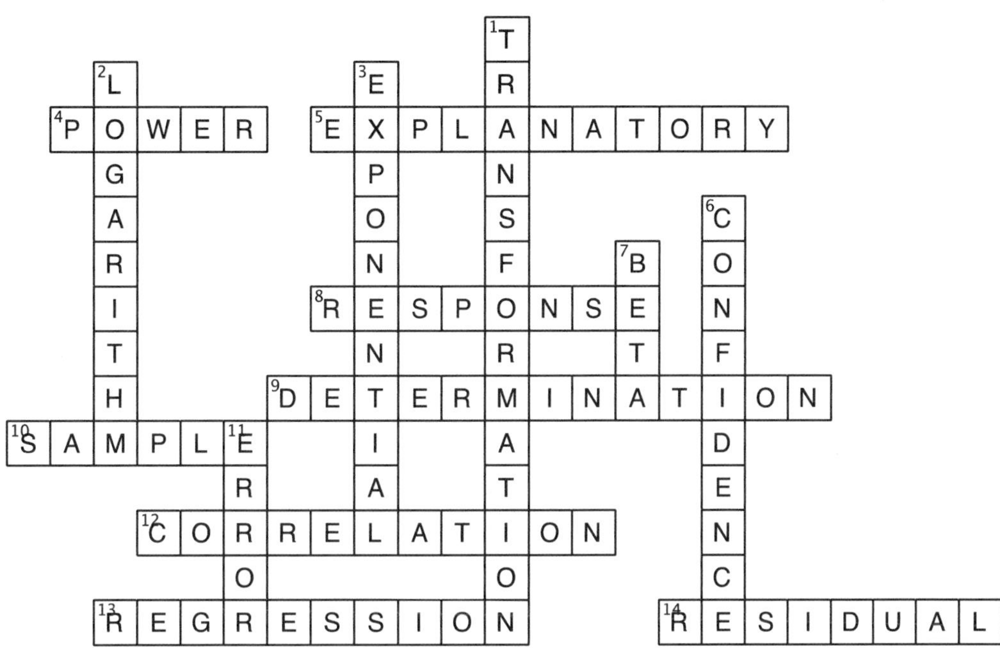


**Across**

4. if (ln(x), ln(y)) is linear, a/an ____ model may be most appropriate for (x,y) [POWER]
5. another name for the x-variable [EXPLANATORY]
8. another name for the y variable [RESPONSE]
9. the coefficient of ____ indicates the fraction of variability in predicted y values that can be explained by the LSRL of y on x [DETERMINATION]
10. when we construct a LSRL for a set of observed data, we construct a ____ regression line [SAMPLE]
12. the _____ coefficient is a measure of the strength of the linear relationship between two quantitative variables [CORRELATION]
13. the line that describes the relationship between an explanatory and response variable [REGRESSION]
14. observed y - predicted y [RESIDUAL]

**Down**

1. when data displays a curved relationship, we can perform a ____ to "straighten" it out [TRANSFORMATION]
2. the common mathematical transformation used in this chapter to achieve linearity for a relationship between two quantitative variables [LOGARITHM]
3. if (x, ln(y)) is linear, a/an ____ model may be most appropriate for (x,y) [EXPONENTIAL]
6. to estimate the population (true) slope, we can construct a ____ interval [CONFIDENCE]
7. the Greek letter we use to represent the slope of the population (true) regression line [BETA]
11. to calculate a confidence interval, we must use the standard ____ of the slope [ERROR]

# Preparing for the AP Statistics Examination

## Overview

After studying and working on statistics all year, you will have a chance to take the AP Statistics Exam in May. Not only will you be able to apply all that you have learned, but you will have an opportunity to earn college credit for your efforts! If you earn a "passing score" on the exam, you may be eligible to receive AP credit at a college or university. This means you will have demonstrated a level of knowledge equivalent to that of students completing an introductory statistics course. You may gain credit hours, advanced placement in a course sequence, and possibly a savings in tuition! For these reasons, it is to your advantage to do your best on the exam. Hopefully this guide has helped you understand the concepts in AP Statistics. Not only is a strong conceptual understanding necessary, but you must also plan and prepare for the exam itself. This section is designed to help you as you get ready for the exam. Best of luck in your studies!

The number of students taking the AP Statistics Exam has been increasing in recent years. In 2011, almost 143,000 students took the exam and earned scores between 1 and 5.

| Score | Qualification | Translation |
|---|---|---|
| 1 | No recommendation | No credit, but you're still better off after having taken the course! |
| 2 | Possibly qualified | Credit? Probably not. |
| 3 | Qualified | Maybe college credit |
| 4 | Well qualified | Most likely college credit |
| 5 | Extremely well qualified | Statistical rock star…woohoo! |

Of the 143,000 students who took the exam in 2011, 58.8% earned a score of 3 or higher and have a good chance of earning college credit. And just think, the majority of students taking the exam probably didn't prepare as much as you did! After working through this Strive Guide along with your textbook, you are well poised to earn a high score on the exam!

## Sections

1. **Sample Schedule**
2. **The AP Statistics Course Outline**
3. **The AP Exam Format**
4. **Planning Your Exam Preparation and Review**
5. **Test Taking Tips**
6. **Practice Exams**

## Section 1 – Sample Schedule

This section presents a general outline to help you prepare and review for the AP Statistics Exam. Every student has their own routine and method of studying. Consider these suggestions as you work through the course and create a plan to maximize your chance of earning a 3 or higher on the exam.

1. ***At the start of your AP Statistics course***
   During the first few weeks of the school year, familiarize yourself with the course outline and AP Statistics Exam format. It has been said that any student can hit a target that is clear and holds still for them. The course outline clearly indicates the topics you will study and the exam format is always the same from year to year. Both the course outline and exam format are presented in the next few sections. Additional information on the exam can be found on the College Board's AP Central website.

2. ***During your AP Statistics course***
   As you work through each chapter, be sure to read the textbook and do all assigned practice problems. Use this guide to help organize your notes, define key terms, and practice important concepts. Each chapter in this guide includes practice multiple choice and free-response questions. Be sure to try all of them and note any concepts that give you difficulty.

3. ***Six weeks before the exam***
   About six weeks before the exam, begin planning for your review and preparation. If you are currently taking more than one AP course, be sure you understand when each exam will be given and plan accordingly.

4. ***Four weeks before the exam***
   In early April, you should be wrapping up your studies in the course. This is a good time to attempt a practice test. Your teacher may provide one and/or you may want to take one of the practice exams in this section. Be sure to note any concepts that give you difficulty so you can tailor your preparation and review for the actual exam. There are still several weeks of class left and plenty of time for you to review, practice, and solidify your understanding of the key concepts in the course.

5. ***The week before the exam***
   The week before the exam, you should be done with your studies and should practice, practice, practice! Take this time to refresh your memory on the key concepts and review the topics that gave you the most trouble during the year. Use another practice test to help you get used to the exam format. Be sure to allow 90 minutes for the multiple choice section and 90 minutes for the free-response section. Use only the formulas and tables provided on the actual exam.

6. ***The day of the exam***
   You have prepared and reviewed as much as possible and are ready for the exam! Make sure you get a good night's sleep, eat a good breakfast the morning of the exam, and check that your calculator is working. Bring several pencils, a clean eraser, and arrive at your exam site early. Good luck!

## Section 2 – The AP Statistics Course Outline

The course outline for AP Statistics is provided below. This outline lists the topics covered in the course and the percentage of the exam devoted to that material. Additional information on the course outline can be found at the College Board's AP Central website.

**Course Content Overview**

The topics for AP Statistics are divided into four major themes: exploratory analysis (20-30% of the exam), planning and conducting a study (10-15% of the exam), probability (20-30% of the exam), and statistical inference (30-40% of the exam).

**Topic Outline**

**I. Exploring Data: Describing patterns and departures from patterns (20%–30%)**

A. Constructing and interpreting graphical displays of distributions of univariate data (dotplot, stemplot, histogram, cumulative frequency plot)
   1. Center and spread
   2. Clusters and gaps
   3. Outliers and other unusual features
   4. Shape

B. Summarizing distributions of univariate data
   1. Measuring center: median, mean
   2. Measuring spread: range, interquartile range, standard deviation
   3. Measuring position: quartiles, percentiles, standardized scores (z-scores)
   4. Using boxplots
   5. The effect of changing units on summary measures

C. Comparing distributions of univariate data (dotplots, back-to-back stemplots, parallel boxplots)
   1. Comparing center and spread: within group, between group variation
   2. Comparing clusters and gaps
   3. Comparing outliers and other unusual features
   4. Comparing shapes

D. Exploring bivariate data
   1. Analyzing patterns in scatterplots
   2. Correlation and linearity
   3. Least-squares regression line
   4. Residual plots, outliers and influential points
   5. Transformations to achieve linearity: logarithmic and power transformations

E . Exploring categorical data
   1. Frequency tables and bar charts
   2. Marginal and joint frequencies for two-way tables
   3. Conditional relative frequencies and association
   4. Comparing distributions using bar charts

**II. Sampling and Experimentation: Planning and conducting a study (10%–15%)**

A. Overview of methods of data collection
   1. Census
   2. Sample survey
   3. Experiment
   4. Observational study

B. Planning and conducting surveys
   1. Characteristics of a well-designed and well-conducted survey
   2. Populations, samples and random selection
   3. Sources of bias in sampling and surveys
   4. Sampling methods, including simple random sampling, stratified random sampling and cluster sampling

C. Planning and conducting experiments
   1. Characteristics of a well-designed and well-conducted experiment
   2. Treatments, control groups, experimental units, random assignments and replication
   3. Sources of bias and confounding, including placebo effect and blinding
   4. Completely randomized design
   5. Randomized block design, including matched pairs design

D. Generalizability of results and types of conclusions that can be drawn from observational studies, experiments and surveys

**III . Anticipating Patterns: Exploring random phenomena using probability and simulation (20%–30%)**

A. Probability
   1. Interpreting probability, including long-run relative frequency interpretation
   2. The Law of Large Numbers
   3. Addition rule, multiplication rule, conditional probability and independence
   4. Discrete random variables and their probability distributions, including binomial and geometric random variables
   5. Simulation of random behavior and probability distributions
   6. Mean (expected value) and standard deviation of a random variable, and linear transformation of a random variable

B. Combining independent random variables
   1. Notion of independence versus dependence
   2. Mean and standard deviation for sums and differences of independent random variables

C. The Normal distribution
   1. Properties of the Normal distribution
   2. Using the Normal distribution table
   3. The Normal distribution as a model for measurements

D. Sampling distributions
   1. Sampling distribution of a sample proportion
   2. Sampling distribution of a sample mean
   3. The Central Limit Theorem
   4. Sampling distribution of a difference between two independent sample proportions
   5. Sampling distribution of a difference between two independent sample means
   6. Simulation of sampling distributions
   7. The *t*-distribution
   8. The Chi-square distribution

**IV. Statistical Inference: Estimating population parameters and testing hypotheses (30%–40%)**

A. Estimation (point estimators and confidence intervals)

1. Estimating population parameters and margins of error
2. Properties of point estimators, including unbiasedness and variability
3. Logic of confidence intervals, meaning of confidence level and confidence intervals, and properties of confidence intervals
4. Large sample confidence interval for a proportion
5. Large sample confidence interval for a difference between two proportions
6. Confidence interval for a mean
7. Confidence interval for a difference between two means (independent and paired samples)
8. Confidence interval for the slope of a least-squares regression line

B. Tests of significance

1. Logic of significance testing, null and alternative hypotheses; p-values; one- and two-sided tests
2. Concepts of Type I and Type II errors; concept of power
3. Large sample test for a proportion
4. Large sample test for a difference between two proportions
5. Test for a mean
6. Test for a difference between two means (independent and paired samples)
7. Chi-square test for goodness of fit, homogeneity of proportions, and independence (one- and two-way tables)
8. Test for the slope of a least-squares regression line

## Section 3 – The AP Statistics Exam Format

The AP Statistics Exam is divided into two sections. The first section consists of 40 multiple-choice questions and the second section consists of 6 free-response questions. Each section counts for 50% of the total exam score. The number of questions you will be asked from each section of the topic outline corresponds to the percentages provided in the AP Statistics Course outline. On the free-response section, you will most likely be asked at least one question about exploratory data analysis, at least one about probability, at least one about sampling or experimental design, and at least one about inference. Question 6 on the free-response section is considered an "investigative task" that will stretch your skills beyond what you have learned in the course and is worth almost twice that of each of the other 5 questions. Even though this question most commonly deals with an unfamiliar topic, you should be able to provide a reasonable answer based on your preparation.

### The Multiple-Choice Section

You will have 90 minutes to complete the 40 question multiple-choice section of the exam. Each question has five answer choices (A-E), only one of which is correct. Each correct answer earns you one point, while each question answered incorrectly (or left blank) earns you no points. It is in your best interest to answer every question, even if you need to make an educated guess! The questions will cover all of the topics from the course and may require some calculation. Some may require interpreting a graph or reflecting on a given situation. Keep in mind you have just over 2 minutes per question. Don't feel it is necessary to work the questions in order. A good strategy is to work all of the ones you feel are easy and then go back to the ones that might take a little more time. Be sure to keep an eye on the clock!

### The Free-Response Section

The second section of the exam is made up of 6 free-response questions including one "investigative task." You will have 90 minutes to complete this section of the exam. You should allow about 25 minutes for the investigative task, leaving just over 12 minutes for each of the other questions. These questions are designed to measure your statistical reasoning and communication skills and are graded on the following 0-4 scale.

4 = Complete Response {NO statistical errors and clear communication}
3 = Substantial Response {Minor statistical error/omission or fuzzy communication}
2 = Developing Response {Important statistical error/omission or lousy communication}
1 = Minimal Response {A "glimmer" of statistical knowledge related to the problem}
0 = Inadequate Response {Statistically dangerous to self and others}

Each problem is graded holistically by the AP Statistics Readers, meaning your entire response to the problem and all its parts is considered before a score is assigned. Be sure to keep an eye on the clock so you can provide at least a basic response to each question!

## Section 4 – Planning Your Exam Preparation and Review

This book has been designed to help you identify your statistical areas of strength and areas in which you need improvement. If you have been working through the checks-for-understanding, multiple-choice questions, FRAPPYs, and vocabulary puzzles for each chapter, you should have an idea which topics are in need of additional study or review. The practice tests that are included in the section of the book are designed to help you get familiar with the format of the exam as well as check your understanding of the key concepts in the course. Plan on taking both of these tests as part of your preparation and review. Allow yourself 90 minutes for each section of the test and be sure to check your answers in the provided keys. For each question you missed, determine whether it was a simple mistake or whether you need to go back and study that topic again. After you complete the tests, continue practicing problems before the exam date. The best preparation for the exam (other than having a solid understanding of statistics) is to practice as many multiple-choice and free-response questions as possible. Ask your teacher or refer to the College Board's AP Central website for additional resources to help you with this!

## Section 5 – Test Taking Tips

Once you have mastered all of the concepts and have built up your statistical communication skills, you are ready to begin reviewing for the actual exam. The following tips were written by your textbook's author, Daren Starnes, and a former AP Statistics teacher, Sanderson Smith, and are used with permission.

### General Advice

**Relax, and take time to think!** Remember that everyone else taking the exam is in a situation identical to yours. Realize that the problems will probably look considerably more complicated than those you have encountered in other math courses. That's because a statistics course is, necessarily, a "wordy" course.

**Read each question carefully before you begin working.** This is especially important for problems with multiple parts or lengthy introductions. Underline key words, phrases, and information as you read the questions.

**Look at graphs and displays carefully**. For graphs, note carefully what is represented on the axes, and be aware of number scale. Some questions that provide tables of numbers and graphs relating to the numbers can be answered simply by "reading" the graphs.

**About graphing calculator use:** As noted throughout this guide, your graphing calculator is meant to be a tool and is to be used sparingly on some exam questions. Your brain is meant to be your primary tool. Don't waste time punching numbers into your calculator unless you're sure it is necessary. Entering lists of numbers into a calculator can be time-consuming, and certainly doesn't represent a display of statistical intelligence. Do not write directions for calculator button-pushing on the exam and avoid calculator syntax, such as *normalcdf* or *1-PropZTest.*

**Multiple-choice questions:**

- Examine the question carefully. What statistical topic is being tested? What is the purpose of the question?
- Read carefully. After deciding on an answer, make sure you haven't made a careless mistake or an incorrect assumption.
- If an answer choice seems "too obvious," think about it. If it's so obvious to you, it's probably

obvious to others, and chances are good that it is not the correct response.

- Since there is no penalty for a wrong answer or skipped problem, it is to your advantage to attempt every question or make an educated guess, if necessary.

**Free-response questions:**

- Do not feel it necessary to work through these problems in order. Question 1 is meant to be straightforward, so you may want to start with it. Then move to another problem that you feel confident about. Whatever you do, don't run out of time before you get to Question 6. This Investigative Task counts almost twice as much as any other question.
- Read each question carefully, sentence by sentence, and underline key words or phrases.
- Decide what statistical concept/idea is being tested. This will help you choose a proper approach to solving the problem.
- You don't have to answer a free-response question in paragraph form. Sometimes an organized set of bullet points or an algebraic process is preferable. NEVER leave “bald answers” or “just numbers” though!
- ALWAYS answer each question in context.
- The amount of space provided on the free-response questions does not necessarily indicate how much you should write.
- If you cannot get an answer to part of a question, make up a plausible answer to use in the remaining parts of the problem.

**On problems where you have to produce a graph:**

- Label and scale your axes! Do not copy a calculator screen verbatim onto the exam.
- Don't refer to a graph on your calculator that you haven't drawn. Transfer it to the exam paper. Remember, the person grading your exam can’t see your calculator!

**Communicate your thinking clearly.**

- Organize your thoughts before you write, just as you would for an English paper.
- Write neatly. The AP Exam Readers cannot score your solution if they can’t read your writing!
- Write efficiently. Say what needs to be said, and move on. Don't ramble.
- The burden of communication is on you. Don't leave it to the reader to make inferences.
- When you finish writing your answer, look back. Does the answer make sense? Did you address the context of the problem?

**Follow directions. If a problem asks you to "explain" or "justify," then be sure to do so.**

- Don't "cast a wide net" by writing down everything you know, because you will be graded on everything you write. If part of your answer is wrong, you will be penalized.
- Don't give parallel solutions. Decide on the best path for your answer, and follow it through to the logical conclusion. Providing multiple solutions to a single question is generally not to your advantage. You will be graded on the lesser of the two solutions. Put another way, if one of your solutions is correct and another is incorrect, your response will be scored "incorrect."

Remember that your exam preparation begins on the first day of your AP Statistics class. Keep in mind the following advice throughout the year.

- READ your statistics book. Most AP Statistics Exam questions start with a paragraph that describes the context of the problem. You need to be able to pick out important statistical cues. The only way you will learn to do that is through hands-on experience.
- PRACTICE writing about your statistical thinking. Your success on the AP Statistics Exam depends not only on how well you can “do” the statistics, but also on how well you explain your reasoning.
- WORK as many problems as you can in the weeks leading up to the exam.

## Section 6 – Practice Exams

Use the following two practice exams to help you prepare for the AP Statistics Exam. Allow yourself 90 minutes for the multiple-choice questions, take a break, and allow yourself 90 minutes for the free-response section. When you finish, correct your exam and note your areas in need of improvement. Answers for each exam are included in the answer key.

## Practice Exam 1

### Section 1: Multiple-Choice

1. To test the effectiveness of a new medicine for the prevention of the common cold, a researcher would like to select 5 subjects from a population of 75 volunteers. The volunteers are labeled 01, 02, 03, ..., 75. Using the following line from a random number table, which of the following represents the sample of 5?

   22368 46573 95225 85393 30995 89198 27982 53401 93965 34095 52666

   (A) 22, 36, 8, 46, 57
   (B) 22, 46, 30, 27, 53
   (C) 22, 36, 65, 73, 58
   (D) 22, 36, 65, 73, 22
   (E) 22, 23, 36, 46, 65

2. In a study to determine whether or not listening to classical music has a positive effect on math performance, 100 students were randomly assigned to two groups. Both groups received the same instruction in mathematics from the same teacher. One group listened to classical music while studying and the other group studied in silence. Both groups were then administered the same 50 point exam. Which of the following procedures would be most appropriate for determining which studying environment was more effective in increasing math performance?

   (A) A *t*-test for the slope of the regression line.
   (B) A matched pairs *t*-test.
   (C) A two-sample *z*-test for proportions.
   (D) A two-sample *t*-test for means.
   (E) A chi-square test for independence.

3. In a recent survey, 850 randomly selected adults asked how often they stopped for coffee on the way to work each week. From the results of the survey, 42.6% indicated that they stopped for coffee at least 3 times per week. Which of the following represents a 95% confidence interval to estimate the true proportion of adults who stop for coffee at least 3 times per week?

   (A) $42.6 \pm 1.960 \cdot \sqrt{\frac{(42.6)(57.4)}{850}}$
   (B) $42.6 \pm 1.960 \cdot \left(\frac{(42.6)(57.4)}{850}\right)$
   (C) $42.6 \pm 1.960 \cdot \sqrt{\frac{(0.426)(0.574)}{850}}$
   (D) $0.426 \pm 1.960 \cdot \sqrt{\frac{(0.426)(0.574)}{849}}$
   (E) $0.426 \pm 1.960 \cdot \sqrt{\frac{(0.426)(0.574)}{850}}$

4. A quality control inspector for a potato chip company would like to estimate the proportion of potatoes in a shipment that meet standards. What sample size should she choose to estimate the true proportion of potatoes that meet standards to within 2% of the true value with 99% confidence?

   (A) 4148
   (B) 339
   (C) 165

(D) 50
(E) 33

5. A school counselor believes that students who enroll in a fine arts class (like band, orchestra, or choir) are more likely to participate in a fine arts activity (such as the school musical, play, or speech). The table below shows data that were collected from a random sample of high school students in a large city.

| | | Class | | | |
|---|---|---|---|---|---|
| | | Band | Orchestra | Choir | Total |
| Activity | One-Act Play | 42 | 21 | 5 | 68 |
| | Musical | 120 | 59 | 10 | 189 |
| | Speech | 20 | 10 | 15 | 45 |
| | Total | 182 | 90 | 30 | 302 |

What is the expected count for the cell for the musical and orchestra?

(A) 59
(B) $\frac{(59)(90)}{302}$
(C) $\frac{(59)(189)}{302}$
(D) $\frac{(90)(189)}{302}$
(E) $\frac{(59)(302)}{189}$

6. A researcher wants to compare the typing speed (in words per minute) using a touchscreen keyboard vs. the speed of typing on a standard size detached keyboard. Which of the following designs would be most effective to test the hypothesis that there is no difference in typing speeds between the keyboard types?

(A) Select 100 people who use computers and randomly divide them into two groups – one that uses the onscreen keyboard and one that uses the detached keyboard. Have each person type the same passage and compare the mean words per minute of each group.
(B) Select 100 people who use computers and have them select the keyboard with which they are most comfortable (onscreen or detached). Have each person type the same passage and compare the mean words per minute of each group.
(C) Find 50 people that use onscreen keyboards regularly and 50 people that prefer using detached keyboards. Have each person type the same passage and compare the mean words per minute of each group.
(D) Select 100 people who use computers and randomly divide them into two groups – one that uses the onscreen keyboard and one that uses the detached keyboard. Observe each person's typing for one day and compare the mean words per minute for each group.
(E) Select 100 people who use computers. Have each person type the same passage using each type of keyboard, randomly selecting which keyboard they will use first. Calculate the difference in words per minute (onscreen – detached) for each person and determine the mean difference.

7. A cereal manufacturer is packaging collectible cards from a popular movie in each box of cereal. One card, a holographic card of a memorable scene, is more rare and occurs in only 3% of boxes. Which of the following represents the probability that the fifth box you purchase will contain your first holographic card?

(A) $({}_5C_1)(.97)^4(.03)$
(B) $(.97)^4(.03)$
(C) $(.97)(.03)^4$
(D) $1-(.97)^5$
(E) $1-(.03)^5$

8. Students in 5 different 2nd grade classes are timed to see how long it takes to correctly answer 50 addition and subtraction problems. Each class was then taught a different strategy for basic arithmetic for one month. After the month, the students were again timed to see how long it takes to correctly answer 50 similar problems. The dot plots below show the differences in times for the students (difference = after – before) for each of the five classes. Which dot plot appears to indicate an improvement in time for the class?

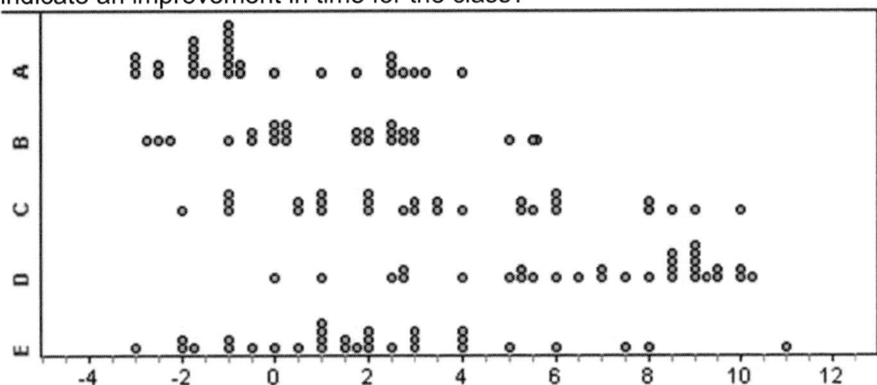


9. Boxplots of two data sets are shown below.

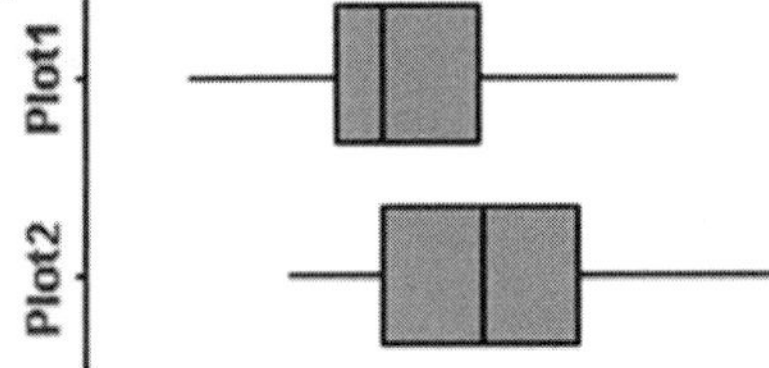


Based on the boxplots, which of the following statements is TRUE?

(A) The spread of both plots is about the same.
(B) The medians of both data sets are approximately equal.
(C) Half of the data values in Plot2 are greater than 75% of the data values in Plot1.
(D) Plot2 contains more data points than Plot1.
(E) Plot1 is more symmetric than Plot2.

10. A researcher has gathered data on the scale scores of students who take the Bush Aptitude in Statistics Exam (BASE) at two northeast schools, Buckley University and Bready College. Boxplots of the score distributions are given below.

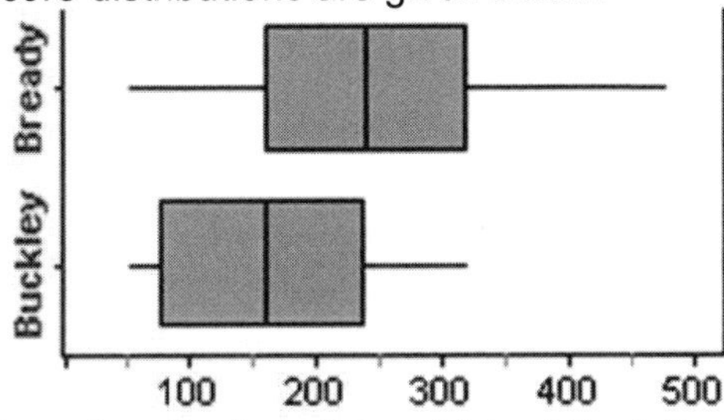


Based on the boxplots, which of the following statements is TRUE?

(A) Since the distributions overlap, there is not much difference between scores at the two schools.
(B) Students at Buckley University have a higher mean score and exhibit more variability in scores.
(C) Students at Buckley University have a higher mean score and exhibit less variability in scores.

(D) Students at Bready College have a higher mean score and exhibit more variability in scores.
(E) Students at Bready College have a higher mean score and exhibit less variability in scores.

11. Researchers would like to know if a simple breathing technique can help relax students who suffer from severe test anxiety. The breathing technique is taught to 42 high school seniors who report they suffer from test anxiety. After a semester of using the technique, 31 of the volunteers report they feel more relaxed during their exams. Which of the following is the main criticism of this design?

(A) Students who do not suffer from severe test anxiety were not included.
(B) No standard measure of anxiety was used.
(C) Since the students volunteered, we cannot conclude the technique was effective.
(D) The researchers should have randomly sampled students from all grades.
(E) No control group was used to account for improvement due to the effect of any type of treatment.

12. A random sample of 83 bicycle enthusiasts found the mean number of miles ridden over the summer is 684 with a standard deviation of 154.5. The 90% confidence interval for the true mean number of miles ridden over the summer for all bicycle enthusiasts is (655.79, 712.21). Which statement below is the correct interpretation of the confidence level?

(A) There is a 90% chance the true mean number of miles ridden is between 655.79 and 712.21.
(B) 90% of all bicycle enthusiasts ride between 655.79 and 712.21 miles over the summer.
(C) We are 90% confident that a randomly selected cyclist will ride between 655.79 and 712.21 miles over the summer.
(D) If 50 random samples of size $n$=83 were taken from the population of bicycle enthusiasts and a 90% confidence interval was calculated for each sample, then the true mean number of miles ridden would be contained within 45 of the calculated intervals, on average.
(E) In repeated sampling, the mean number of miles ridden over the summer will be contained within the interval 90% of the time.

13. Biologists in Minnesota are interested in determining if there is a difference in the invasion rate of Asian Carp (which can be detrimental to the environment) between the Mississippi River and Lake Mille Lacs. In the Mississippi River, it was found that 206 of 579 fish caught were Asian Carp. In Lake Mille Lacs, 28 of 132 fish caught were Asian Carp. Let $p_1$ = the true proportion of Asian Carp in the Mississippi River and let $p_2$ = the true proportion of Asian Carp in Lake Mille Lacs. A test of $H_0$: $p_1 = p_2$ vs. $H_a$: $p_1 \neq p_2$ resulted in a $p$-value of 0.0015. Which of the following is a correct conclusion?

(A) The test is not appropriate, since the researchers should have conducted a one-sided test.
(B) The test is not appropriate, since the sample size is too small to conduct an inference test for proportions.
(C) The test is not appropriate, since the two sample sizes are very different.
(D) The $p$-value of this test is large, indicating we have sufficient evidence to conclude that a difference exists in the invasion rates of Asian Carp for the Mississippi River and Lake Mille Lacs.
(E) The $p$-value of this test is smaller than most reasonable significance levels, indicating we have sufficient evidence to conclude that a difference exists in the invasion rates of Asian Carp between the Mississippi River and Lake Mille Lacs.

14. To determine whether or not a new fertilizer is effective in increasing the yield of tomato plants, researchers planted 140 tomato plants in a research field. A random selection of 60 plants received the new fertilizer, 40 received a traditional fertilizer, and 40 received no fertilizer. What are the experimental units in this study?

(A) The 140 plants
(B) The 60 plants treated with the new fertilizer
(C) The 40 plants treated with the traditional fertilizer
(D) The 40 plants that were not treated

(E) The population of all tomato plants

15. The random variable $X$ has a continuous uniform probability distribution shown below.

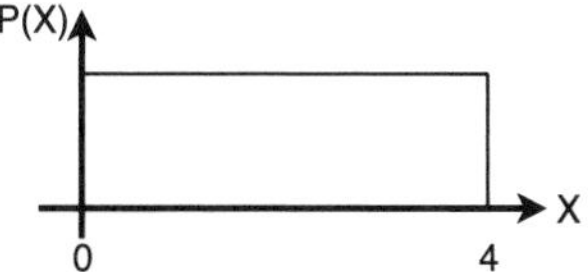


What is $P(X \leq 0.7)$?

(A) 0.825
(B) 0.7
(C) 0.3
(D) 0.25
(E) 0.175

16. The amount of money carried by boys at Lakeville South High School is Normally distributed with a mean \$32 and standard deviation \$6. The amount of money carried by girls at the same high school is Normally distributed with a mean \$23 and standard deviation \$4. Suppose you randomly select one boy and one girl from this school. Assuming the amounts of money carried by each are independent, what is the approximate probability that the girl's amount is at least \$5 less than the boy's amount?

(A) 0.20
(B) 0.29
(C) 0.34
(D) 0.55
(E) 0.71

17. While doing her homework, a student constructed a least-squares regression line for a set of data and found the correlation coefficient to be $r = 0.68$. She then realized she switched the explanatory and response variables. After constructing a new least-squares regression line, what will the correlation be?

(A) 0.68
(B) 0.4624
(C) 1 / 0.68
(D) -1 / 0.68
(E) -0.68

18. A recent study of 1000 high school juniors and seniors asked whether or not they felt smartphones are a valuable learning tool. Additional data were gathered on their smartphone plan (unlimited data, additional texting, or standard voice plan) and which class they were in (junior vs. senior). Which of the following is a correct pair of hypotheses the researchers could test?

(A) $H_0$: The proportion of juniors and seniors who feel smartphones are a valuable learning tool is the same for students that have each type of plan.
$H_a$: The proportion of juniors and seniors who feel smartphones are a valuable learning tool is not the same for students that have each type of plan
(B) $H_0$: The proportion of students who feel smartphones are a valuable learning tool is the same for students that have each type of plan.
$H_a$: The proportion of students who feel smartphones are a valuable learning tool is not the same for students that have each type of plan.
(C) $H_0$: There is an association between class and type of plan.
$H_a$: There is no association between class and type of plan.

(D) $H_0$: Class and type of plan are not independent.
$H_a$: Class and type of plan are independent.
(E) $H_0$: Class and type of plan are independent.
$H_a$: Class and type of plan are not independent.

19. Four distributions are shown below. One represents the distribution of the values in the population. One represents a sampling distribution of means for samples of size $n = 2$, $n = 10$, and $n = 25$. What is the correct order of population, means for $n = 2$, means for $n = 10$, and means for $n = 25$?

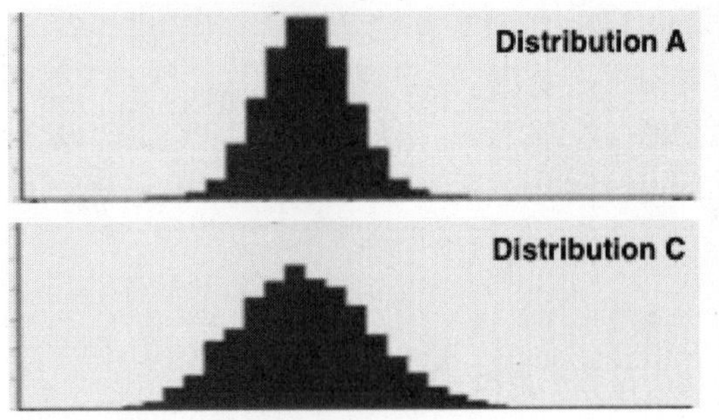

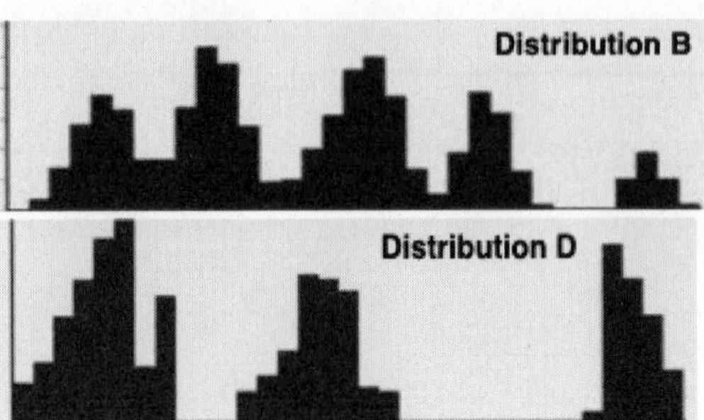


(A) A, B, C, D
(B) D, B, C, A
(C) B, D, C, A
(D) A, C, B, D
(E) C, A, B, D

20. The following data were gathered as part of a study to determine educational levels for different professions.

| Degree | Bachelor's | Master's | Professional | Doctorate | Other | Total |
|---|---|---|---|---|---|---|
| Educator | 12 | 23 | 42 | 45 | 18 | 140 |
| Researcher | 15 | 25 | 51 | 41 | 18 | 150 |
| Biologist | 32 | 38 | 31 | 17 | 12 | 130 |
| Author | 17 | 15 | 10 | 8 | 7 | 57 |
| Total | 76 | 101 | 134 | 111 | 55 | 477 |

What is the probability that a randomly chosen person from this group is a researcher, given that the person has a doctorate?

(A) less than 0.001
(B) 0.09
(C) 0.31
(D) 0.37
(E) 0.74

21. Two drugs are available to treat the flu. Adiflu is effective in curing 80% of flu cases, but results in unpleasant side effects for 45% of individuals who take it. Beneflu cures 55% of flu cases but results in unpleasant side effects for 25% of individuals who take it. Suppose 70 people take Adiflu and 90 people take Beneflu. What is the expected number of people who will experience unpleasant side effects?

(A) 37.575
(B) 54
(C) 58
(D) 105.5
(E) 110.5

22. An experiment was conducted to determine the time it takes students to complete a basic arithmetic test on a computer compared to using paper and pencil. Students took two similar arithmetic tests,

one on a computer and one using paper and pencil. Students were randomly assigned the order of the tests. The top plot shows the times to complete the test using paper and pencil. The bottom plot shows the times to complete the test on a computer.

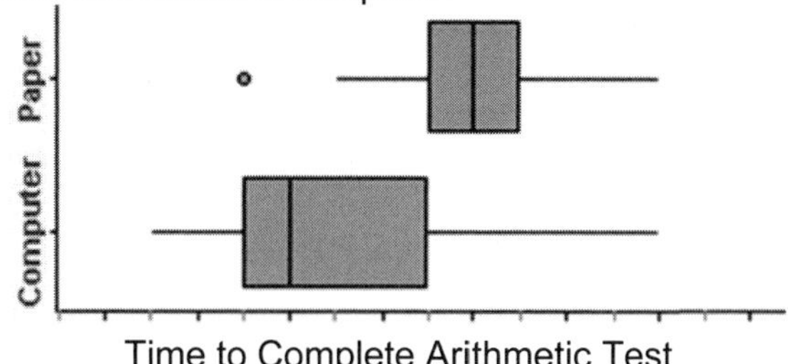


Which of the following statements is correct?

(A) About 90% of tests taken on the computer were completed faster than those on paper.
(B) The standard deviation of the paper tests is larger than the standard deviation of the tests taken on the computer.
(C) There were more tests taken on the computer than paper and pencil.
(D) The times for the tests taken on the computer would result in a narrower 90% confidence interval than the tests taken on paper.
(E) About 75% of the tests taken on the computer were completed faster than all but 25% of the paper tests.

23. A poll of 83 AP Statistics students asked, "How many hours a week do you spend studying statistics?" The results are summarized in the computer output below.

| Variable | N | Mean | | Median | TrMean | StDev | |
|---|---|---|---|---|---|---|---|
| Hours | | 83 8.5 | | 7.2 | | 73.8 | 4.9 |
| | | Minimum | Maximum | | Q1 | Q3 | |
| Hours | | 0.00 | 25.00 | | 5 | 12 | |

If a boxplot of this data was constructed, what data points would be marked as outliers?

(A) Any points greater than 10.5
(B) Any points greater than 17
(C) Any points greater than 17.7
(D) Any points greater than 22.5
(E) Any points greater than 25

24. A veterinarian was interested in comparing the effects of two drugs, glucosamine and chondroitin, on arthritis in beagles. She believes glucosamine is more effective in increasing mobility for dogs that suffer from arthritis in their hips. A sample of 150 beagles showed no difference in mobility between dogs treated with each drug. When the sample size was increased to 475 beagles, the results showed that glucosamine was significantly more effective in increasing mobility. This is an example of

(A) smaller sample sizes leading to lower power.
(B) using a test for means instead of a test for proportions.
(C) reducing the probability of a Type I error.
(D) using blocking to reduce variation.
(E) always using large samples.

25. It is believed that positive, self-affirming statements can lead to greater academic success on high-stakes assessments. In a recent study, 120 advanced algebra students were randomly assigned to two groups. One group was instructed how to write and recite daily positive self-affirmations while the other group did not do any affirmations. After one semester, all 120 students took the same algebra examination. Which of the following would be appropriate hypotheses for this study?

(A) $H_o$: $\mu_1 = \mu_2$ and $H_a$: $\mu_1 < \mu_2$, where $\mu_1$ = mean score for those who used affirmations and $\mu_2$ = mean score for those who don't use affirmations.
(B) $H_o$: $\mu_1 = \mu_2$ and $H_a$: $\mu_1 \neq \mu_2$, where $\mu_1$ = mean score for those who used affirmations and $\mu_2$ = mean score for those who don't use affirmations.
(C) $H_o$: $\mu_1 = \mu_2$ and $H_a$: $\mu_1 > \mu_2$, where $\mu_1$ = mean score for those who used affirmations and $\mu_2$ = mean score for those who don't use affirmations.
(D) $H_o$: $\mu_1 > \mu_2$ and $H_a$: $\mu_1 = \mu_2$, where $\mu_1$ = mean score for those who used affirmations and $\mu_2$ = mean score for those who don't use affirmations.
(E) $H_o$: $\mu_1 \neq \mu_2$ and $H_a$: $\mu_1 = \mu_2$, where $\mu_1$ = mean score for those who used affirmations and $\mu_2$ = mean score for those who don't use affirmations.

26. Suppose the mean score on a recent standardized test was 3.20. A student who earned a score of 4.10 was in the 85$^{th}$ percentile. If the scores are normally distributed, what is the approximate standard deviation of the scores on the standardized test?

(A) 1.04
(B) 0.90
(C) 0.87
(D) 1.15
(E) 1.28

27. Which of the following statements about confidence intervals is TRUE?

(A) Changing the confidence level does not affect the width of the interval.
(B) As the sample size increases, the width of the interval increases.
(C) Confidence intervals cannot be used to make decisions for a test of significance.
(D) Larger sample means result in wider confidence intervals.
(E) The width of a confidence interval is affected by the variation in the original population.

28. A shoe manufacturer claims their new hiking boot is more resistant to wear and tear than their current hiking boot on the market. To test this claim, 27 hikers agree to wear one of the new boots on one foot and one of the current boots on the other foot for the next hiking season. The foot that will wear the new boot is decided by a coin flip for each hiker and the current boot is placed on the other foot. After a full season of hiking, the boots are collected and the differences in wear between the two boots are calculated for each hiker. Which of the following is the main reason for conducting a matched pairs test in this case?

(A) If each hiker wears both boots, a smaller sample size ($n < 30$) will suffice.
(B) Variation associated with lurking variables will be reduced, since each hiker acts as a control.
(C) Since there is only one parameter (difference), we can check fewer conditions.
(D) Randomizing which foot gets which boot eliminates all bias.
(E) The $t$ test statistic is more accurate because we are dealing with fewer values.

29. To test the effectiveness of a new pesticide, 24 plots in a cornfield are set up. 8 randomly selected plots are treated with the new pesticide. Another 8 plots are randomly selected and treated with the traditional pesticide. The remaining 8 plots receive no pesticide. What is the purpose of randomly selecting plots?

(A) Randomization eliminates bias.

(B) Randomization guarantees a more accurate experiment.
(C) Agricultural experiments require randomization.
(D) Randomizing reduces variation due to differences in the plots of land by creating roughly equivalent groups.
(E) Since fewer than 30 plots are selected, randomization is required to meet the conditions of the significance test.

30. A study performed by a psychologist determined that a person's sense of humor is linearly related to their IQ. The equation of the least squares regression line is $\widehat{humor} = -49 + 1.8(IQ)$. What is the residual for an individual with an IQ score of 110 and a humor score of 140?

(A) -30
(B) -9
(C) 9
(D) 30
(E) Cannot be determined since we don't know the original data points.

31. A large high school offers AP Statistics and AP Calculus. Among the seniors in this school, 65% take AP Statistics, 45% take AP Calculus, and 30% take both. If a senior class student is randomly selected, what is the probability they are in AP Statistics or AP Calculus, but not both?

(A) 10%
(B) 35%
(C) 50%
(D) 70%
(E) 75%

32. Which of the following statements is TRUE for the *t*-distribution?

(A) The *t*-distributions and the Normal distribution are not related.
(B) The tails of the *t*-distribution are "lighter" than the tails of the Normal distribution.
(C) The *t*-distribution approaches the Normal distribution as the sample size increases.
(D) The Normal distribution approaches the *t*-distribution as the sample size increases.
(E) The *t*-distribution is more skewed than the Normal distribution.

33. A study is conducted to determine the effectiveness of mental exercises on short-term memory. A randomized comparative experiment is conducted and the number of items recalled on a short-term memory test is recorded for 25 individuals who practiced mental exercises and 35 individuals who did not. The individuals who practiced mental exercises recalled 16.0 items on average with a variance of 6.25. The individuals that did not do any mental exercises recalled 12.5 items on average with a variance of 7.29. Assuming the population of items recalled is Normal for each group, what is the standard deviation of the sampling distribution of the difference in mean items recalled for the two groups?

(A) $\sqrt{6.25 + 7.29}$
(B) $\sqrt{\frac{6.25}{25} + \frac{7.29}{35}}$
(C) $\sqrt{6.25 - 7.29}$
(D) $\sqrt{\frac{6.25}{25} - \frac{7.29}{35}}$
(E) $\sqrt{\frac{6.25}{60} + \frac{7.29}{60}}$

34. A quality control specialist samples 50 bottles of soda labeled “20 ounces” and constructed a 90% confidence interval for the true mean contents of the bottles. The resulting interval was (19.78, 20.82). If the specialist constructed a 95% confidence interval, the result would be

(A) narrower and have a smaller chance of not capturing the true mean.
(B) narrower and have a larger chance of not capturing the true mean.
(C) wider and have a smaller chance of not capturing the true mean.
(D) wider and have a larger chance of not capturing the true mean.
(E) wider, but the chance of not capturing the true mean cannot be determined.

35. According to a recent national survey, 49% of students who attend a public university graduate with at least $15,000 of debt. At a large Midwestern university, a random sample of 200 recent graduates found that 110 of them had at least $15,000 of debt. If the percent of students who carry debt at this university is actually the same as the national average, which of the following represents the probability of getting a sample of 200 graduates in which 55% or more carry at least $15,000 of debt?

(A) $P\left(z > \dfrac{.55-.49}{\sqrt{\dfrac{(.49)(.51)}{200}}}\right)$

(B) $P\left(z > \dfrac{.55-.49}{\sqrt{\dfrac{(.55)(.45)}{200}}}\right)$

(C) $P\left(z > \dfrac{.49-.55}{\sqrt{\dfrac{(.55)(.45)}{200}}}\right)$

(D) $\dbinom{200}{110}(.49)^{110}(.51)^{90}$

(E) $\dbinom{200}{110}(.51)^{110}(.49)^{90}$

36. Consider the scatterplot below:

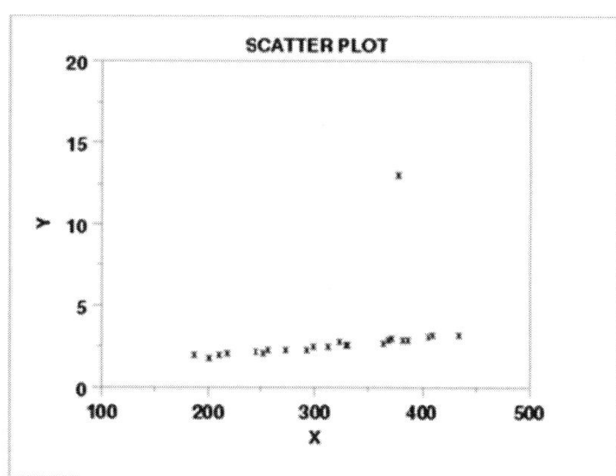


If the outlier (375, 13) is removed, which of the following statements would be TRUE?

(A) The slope of the least-squares regression line would increase and the correlation would decrease.
(B) The slope of the least-squares regression line would increase and the correlation would increase.
(C) The slope of the least-squares regression line would decrease and the correlation would decrease.
(D) The slope of the least-squares regression line would decrease and the correlation would increase.
(E) The slope of the least-squares regression line and the correlation would stay the same.

37. A study of the effect of a fertilizer on the yield of 10 plots of tomato plants resulted in the following computer output:

The regression equation is
Yield = 10.1 + 1.15 Fertilize

| Predictor | Coef | SE Coef | T | P |
|---|---|---|---|---|
| Constant | 10.100 | 0.7973 | 12.67 | 0.000 |
| Fertilize | 1.150 | 0.1879 | 6.12 | 0.000 |

S = 0.8404     R-Sq = 82.4%  R-Sq(adj) = 80.2%

Which of the following would represent a 99% confidence interval to estimate the true slope of the regression line relating amount of fertilizer to yield? Assume all conditions for inference are met.

(A) $1.150 \pm 3.355(0.8404)$

(B) $1.150 \pm 3.169\left(\dfrac{0.8404}{\sqrt{10}}\right)$

(C) $1.150 \pm 6.12(0.1879)$

(D) $1.150 \pm 3.355(0.1879)$

(E) $1.150 \pm 3.169(0.1879)$

38. A poll was conducted to determine the level of support for a candidate in a local election. In a sample of 250 randomly selected adults, 120 indicated their intent to vote for the candidate in the upcoming election. The poll had a margin of error of 3%. What is the correct interpretation of the margin of error?

(A) Approximately 3% of those surveyed refused to answer the question.
(B) The true proportion of adults who would vote for the candidate is likely to be within 3% of the sample proportion.

(C) The true proportion of adults who would vote for the candidate is likely to be at least 3% more than or less than the sample proportion.
(D) If we repeatedly sampled 250 adults at random, the results would vary by no more than 3% from the first sample proportion of 48%.
(E) If we increased the sample size, the proportion of adults who would vote for the candidate would increase by about 3%.

39. A preschool class consists of 15 students ranging in different ages from 2.25 to 5.5 years old. Suppose the youngest child in the class moves to another school and is replaced by a new student who is 6-years old. Which of the following statements is FALSE?

(A) The mean age of the class will increase.
(B) The mode will remain the same.
(C) The standard deviation might be different.
(D) The range of the ages might be different.
(E) The median age will remain the same.

40. A study of the fuel efficiency of 50 car models resulted in the following findings regarding the relationship between the weight of a car (in 1000s of pounds) and its fuel efficiency (in miles per gallon, mpg).

Dependent variable: MPG

| Variable | DF | Parameter Estimate | Standard Error | T for H0: Parameter = 0 | Prob > \|T\| |
|---|---|---|---|---|---|
| INTERCEPT | 1 | 48.7393 | 1.976 | 24.7 | 0.000 |
| Weight | 1 | -8.21362 | 0.6738 | -12.2 | 0.000 |

Based on the information provided, which of the following statements is the best interpretation for the slope of the least-squares regression line?

(A) For each additional 1000 pounds in car weight, the fuel efficiency will decrease by approximately 48.7 mpg.
(B) For each additional 1000 pounds in car weight, the fuel efficiency will decrease by approximately 0.6738 mpg.
(C) For each additional 1000 pounds in car weight, the fuel efficiency will decrease by approximately 8.2 mpg.
(D) For each additional 1000 pounds in car weight, the fuel efficiency will decrease by approximately 12.2 mpg.
(E) For each additional 1000 pounds in car weight, the fuel efficiency will decrease by approximately 1.976 mpg.

## Section 2: Free-Response

1. Fast food restaurants offer convenient meals at low prices. However, while these meals may offer a lot of food at a low price, they can also be high in calories, fat, and sodium. A recent study of 32 fast food items measured their protein content (grams) and fat content (grams). A scatterplot, residual plot, and regression analysis of the observed items are given below.

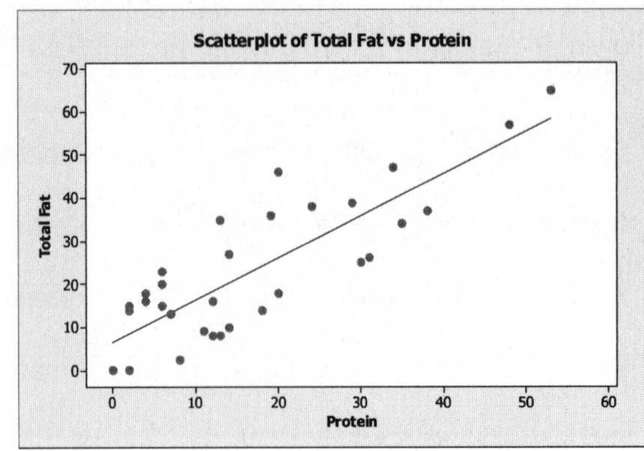


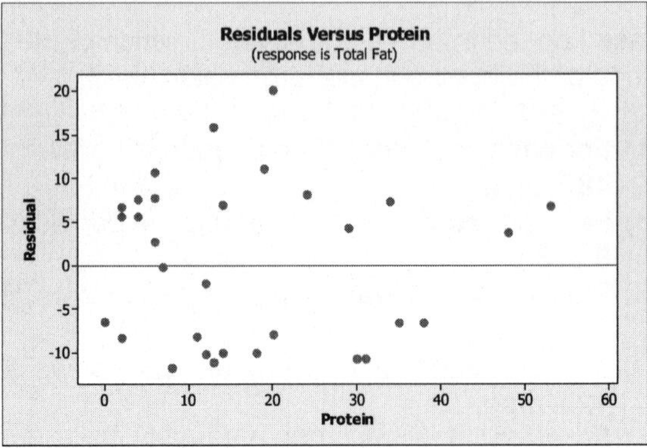


| Predictor | Coef | SE Coef | T | P |
|---|---|---|---|---|
| Constant | 6.411 | 2.6640 | 2.56 | 0.0158 |
| Protein | 0.977 | 0.1213 | 8.05 | 0.0000 |

S = 9.311   R-Sq = 68.39%

(a) Describe the relationship between the grams of protein and grams of fat.

(b) Is it reasonable to use a linear model to describe this relationship? Explain your reasoning.

(c) Write the equation of the least-squares regression line relating grams of protein and grams of fat. Define any variables used.

(d) Identify and interpret the slope of the least-squares regression line.

(e) Interpret the value of *s* in the context of the problem.

2. A pharmaceutical company would like to test the effectiveness of two new influenza vaccines in the prevention of the common flu strain. A group of 600 adult volunteers (ages 25-50) are recruited in a metropolitan area that has a history of high flu rates during the winter months. The volunteers are randomly divided into three groups of 200 individuals each. One group is administered a new injection vaccine, one group is administered a new mist vaccine, and the third group is administered the current vaccine. Researchers will record the number of volunteers who develop the flu in each group.

(a) Is this an experiment or an observational study? Why?

(b) Suppose the company wishes to determine if there is a difference in the effectiveness of the new injection vaccine and the effectiveness of the new mist vaccine. Identify the hypotheses the company would test.

(c) Why would the company want to include a group of volunteers who are administered the current vaccine?

(d) Suppose significant evidence is found that suggests the new vaccines are more effective in reducing the rate of contracting the flu. Can this conclusion be used to say that the new vaccine is more effective in flu-prevention for everyone? Explain.

3. In a recent survey, a random sample of 996 adults was asked whether or not they supported a particular grassroots political movement. The poll found that 558 of those surveyed supported the movement.

(a) Construct and interpret a 95% confidence interval for the true proportion of adults who support the movement.

(b) Interpret the confidence level in the context of the problem.

4. An agricultural researcher measured the yield of tomato plants that were sprayed with two different fertilizers. Ten randomly selected plots were sprayed with "crop increasing fertilizer" (CIF) and 10 randomly selected plots were sprayed with "yield enhancing cultivar" (YEC). The yield was measured in kilograms per hectare (kg/ha). Boxplots and summaries of the data are shown below:

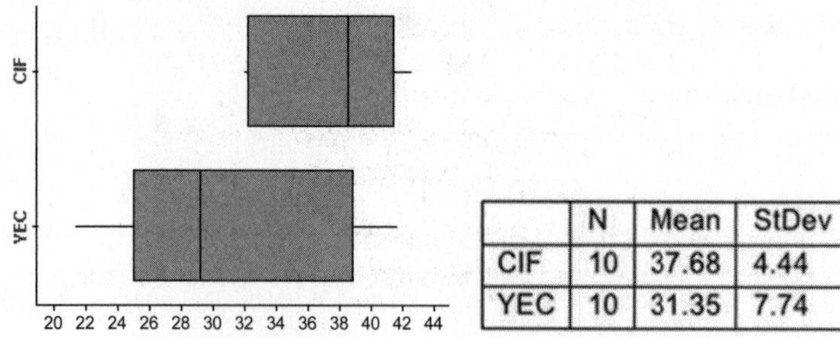


| | N | Mean | StDev |
|---|---|---|---|
| CIF | 10 | 37.68 | 4.44 |
| YEC | 10 | 31.35 | 7.74 |

Is there evidence that CIF is more effective in increasing tomato yield than YEC? Give appropriate statistical evidence to support your answer.

5. An advertising firm offers a promotional package that includes four coffee mugs and five desk calendars, each printed with the logo of your company and shipped in a display box. The manufacturing process of the mugs results in products that have a mean weight of 6 ounces with a standard deviation of 0.2 ounces. The calendars are manufactured with a mean weight of 8 ounces with a standard deviation of 0.3 ounces. The shipping boxes have a mean weight of 3 ounces and a standard deviation of 0.1 ounce.

(a) Assume the weights of the items are independent and normally distributed. What are the mean and standard deviation of the weight of the promotional package?

(b) The firm is charged additional postage for packages that weigh more than 68 ounces. What is the probability that a randomly selected package will require additional postage?

(c) If 4 packages are selected at random, what is the probability that their mean weight is greater than 68 ounces?

6. Nokomis Airline operates a commuter jet that flies between Northfield, MN, and Sheboygan, WI, on a daily basis. The jet has 10 seats for passengers. It is assumed that 90% of passengers show up for each flight while 10% are "no-shows." In an effort to maximize profits and operate a full flight each day, the airline sells 12 tickets per flight.

(a) Describe the design of a simulation using the table of random digits below to estimate the probability a randomly selected flight will be overbooked.

(b) Perform 15 repetitions of your simulation. Clearly indicate your procedure on the table below and use your results to determine the probability a flight will be overbooked.

19223 95034 05756 28713 96409 12531 42544 82853 68417 35013 15529 47487

73676 47150 99400 01927 27754 42648 82425 34290 82739 57890 20807 18883

45467 71709 77558 00095 32863 29485 82226 90056 60940 72024 17868 41979

61041 77684 94322 24709 73689 14526 31893 32592 99278 19931 36809 71868

(c) The histogram below represents the number of no-shows for two years' worth of ticket sales for this daily flight. Does this histogram support the assumption that 90% of passengers will show up for the flight? Assume 12 tickets were sold for each 10-seat flight. Explain your reasoning.

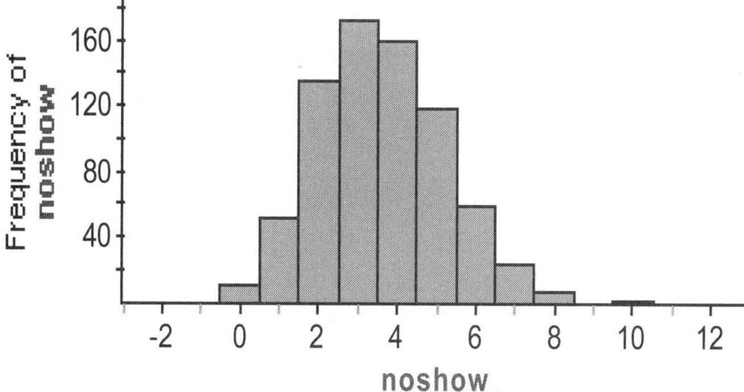

**Answer Key**
**Practice Exam 1**
**Multiple-Choice**

1. 22 368 465 73 95225 85393 30995 89198 27982 53401 93965 34095 52666
   The numbers selected are 22, 36, 65, 73, 58 Ans: C

2. We need to determine if the group that listened to classical music had a higher average score on the exam than those who listened in silence. This would be a two-sample *t*-test. Ans: D

3. To construct a 99% confidence interval, we use $\hat{p} \pm z^* \cdot \sqrt{\frac{\hat{p}(1-\hat{p})}{n}}$. The sample proportion is 0.426 and at 95% confidence, $z^*=1.960$. Ans: C

4. $ME = z^* \sqrt{\frac{p(1-p)}{n}} \Rightarrow .02 = 2.576\sqrt{\frac{.5(.5)}{n}} \Rightarrow n = \frac{0.25}{\left(\frac{.02}{2.576}\right)^2} = 4147.36$ Ans: A

5. Expected (musical and orchestra) = (row total)(column total)/(grand total) = (90)(189)/302
   Ans: D

6. For maximum control, each person should type the same passage using each keyboard. This would minimize differences between experience, preference, and ability. Ans: E

7. Let $X$ = the number of cereal boxes purchased until the holographic card is found. $X$ is a geometric random variable. P(holographic card) = 0.03. We need 4 failures, then a success. Ans: B

8. An increase in speed would mean a smaller "after" time than "before." Since the differences are measured after – before, most of the differences should be negative. Ans: A

9. The median for Plot2 is approximately equal to Q3 for Plot1. Ans: C

10. The boxplot for Bready college is wider than the boxplot for Buckley University, so those scores have more variability. Since both boxplots are fairly symmetric, the mean will be near the median, therefore the mean of scores at Bready College will be higher. Ans: D

11. A treatment was imposed, but a control group should have been set up for comparison purposes. We don't know if the change in anxiety is due to the technique or another change in the environment experienced by all students. Ans: E

12. The confidence level refers to the "success rate" for the method used to construct the interval. We expect 90% of the intervals constructed to contain the true mean. Ans: D

13. The *p*-value is small, indicating we reject the null hypothesis for most reasonable significance levels.
    Ans: E

14. The experimental units are the elements to which treatments are applied. Ans: A

15. The area of the probability distribution must be 1, so the height of the distribution is 0.25.
$P(X \le 0.7)$ = area of the rectangle = 0.7 * 0.25 = 0.175. Ans: E

16. The distribution of (Male amount – Female amount) will be Normally distributed with mean = 32 – 23 = 9 and standard deviation = $\sqrt{(6^2 + 4^2)} = 7.211$. P(diff ≤ 5) = $P(z \le (5-9)/7.211) = 0.2912$.
    Ans: B

17. Correlation does not rely on a distinction between explanatory and response variables. It will remain unchanged when the new regression line is constructed. Ans: A

18. The researchers are interested in determining whether or not the proportion of juniors and seniors who believe smartphones are a valuable learning tool is the same across all types of smartphone plans. Ans: A

19. As the sample size increases, the sampling distribution of sample means will become more Normally distributed and exhibit less variability. Ans: B

20. The conditional probability P(researcher | doctorate) = 41/111 = 0.37. Ans: D

21. Expected number of people with unpleasant side effects = .45(70) + .25(90) = 54 Ans: B

22. The Q1 of the paper tests is equal to Q3 of the computer tests. About 75% of the times taken on the computer test were less (faster) than the Q1of the paper tests. Ans: E

23. Using the 1.5 IQR Rule, an outlier is any value that falls at least 1.5xIQR above Q3 or at least 1.5xIQR below Q1. IQR = 12 – 5 = 7. Q1 – 1.5(7) = -5.5. Q3 + 1.5(7) = 22.5 Ans: D

24. The power of a significance test is its ability to detect a difference. The higher the power, the better the test's ability to detect a significant result. One way to increase power is to increase the sample size. Small sample sizes will have lower power. Ans: A

25. The most appropriate measure would be the average score on the exam. Since it is believed affirmations increase performance, the alternative hypothesis would claim a higher mean score for students who use affirmations than that of students who don't. Ans: C

26. The *z*-score corresponding to the 85th percentile is approximately 1.04.

$$z = \frac{x-\mu}{\sigma} \Rightarrow 1.04 = \frac{4.1-3.2}{\sigma} \Rightarrow \sigma \frac{0.9}{1.04} = 0.8654$$

Ans: C

27. The more variation that exists in the population, the larger the margin of error. Ans: E

28. Variation due to the individual hikers is greatly reduced since each hiker wears both boots. Differences in preferred terrain, hiking habits, etc. are considered in the design. Ans: B

29. Randomization reduces variability due to unaccounted for factors by ensuring equivalent treatment groups. Ans: D

30. Predicted humor score = -49 + 1.8(110) = 149. Residual = observed – predicted = 140 – 149 = -9. Ans: B

31. Since 30% of students take both (65-30)% take Statistics and (45-30)% take Calculus. 35% take Statistics + 15% take Calculus = 50% in one or the other, but not both. Ans: C

32. As sample sizes increase, the t-distribution approaches the Normal distribution. Ans: C

33. The standard deviation of the difference between two means is given by $\sqrt{\frac{\text{var}\,1}{n1}+\frac{\text{var}\,2}{n2}}$

Ans: B

34. A higher confidence level results in a wider confidence interval. However, we cannot calculate the probability of capturing the mean in the interval. Ans: E

35. The sampling distribution of $\hat{p}$ is approximately Normal with mean of p = 0.49 and standard deviation $\sqrt{\frac{(0.49)(0.51)}{200}}$. Therefore, therefore $P\left(z > \frac{\hat{p}-p}{\sqrt{\frac{p(1-p)}{n}}}\right) = P\left(z > \frac{.55-.49}{\sqrt{\frac{(.49)(.51)}{200}}}\right)$ Ans: A

36. The point (375, 13) is influential and "pulls" the LSRL towards it. Removing it would decrease the slope of the LSRL and would increase the correlation. Ans: D

37. The confidence interval is given by $b_1 \pm t^* \cdot SE(b_1)$ with df = 10-2 = 8. Ans: D

38. The margin of error tells us how much the sample statistic and true parameter are likely to differ. Ans: B

39. The median position will shift up one if the children are arranged from youngest to oldest. Ans: E

40. The slope of the least squares regression line is -8.21362. Every increase of one unit for x will result in an average decrease of -8.21362 in y. Ans: C

**Answer Key**
**Practice Exam 1**
**Free Response**

1. (a) There is a moderate/strong, positive, linear relationship between the number of grams of protein and the number of grams of fat for fast food items.

   (b) Yes, it is reasonable to use a linear model for this relationship. The scatterplot suggests a linear relationship exists and the residual plot shows no obvious pattern. Further, the computer output indicates the t-test for slope has a p-value of approximately 0, indicating significant evidence of a positive slope.

   (c) $\widehat{fat} = 6.411 + 0.977(protein)$

   (d) slope = 0.977. For every increase of one gram of protein, the predicted amount of fat increases by approximately 0.977 grams.

   (e) s=9.311. On average, the difference between the predicted amount of fat and actual amount of fat is 9.311 grams.

2. (a) This is an experiment since treatments were imposed on randomly assigned groups of subjects.

   (b) Let $p_i$ = true proportion of people who get the flu after having the injection vaccine and $p_m$ = true proportion of people who get the flu after having the mist vaccine.

   $H_0$: $p_i = p_m$ vs. $H_a$: $p_i \neq p_m$

   (c) Since they want to test the effectiveness of the new vaccine, they may want to compare it to the effectiveness of the old vaccine. While they may find one form (injection or mist) to be more effective than the other, it may be that neither is more effective than the original vaccine.

   (d) No. Only adults were used in the study. The vaccine may not work as well on children or seniors.

3. (a) Let $p$ = the true proportion of adults who support the movement. Since we have a random sample, 996(.56) = 558 and 996(.44)=438 are both greater than 10, and there are more than 9960 adults in the US we can construct a 95% confidence interval.

$$95\%\ CI = \hat{p} \pm z * \sqrt{\frac{\hat{p}(1-\hat{p})}{n}} \Rightarrow 0.56 \pm 1.96\sqrt{\frac{0.56(1-0.56)}{996}} \Rightarrow (0.529, 0.591)$$

We are 95% confident the interval from 0.529 to 0.591 captures the true proportion of adults who support the political movement.

(b) The confidence level is 95%. If we were to repeatedly sample 996 adults at random and construct a confidence interval from each sample, 95% of the intervals constructed would capture the true proportion of adults who support the political movement.

4. Let $\mu_C$ = the true mean yield for tomato plants sprayed with CIF and $\mu_Y$ = the true mean yield for tomato plants sprayed with YEC. We will perform a two-sample *t*-test on the hypotheses:

$H_0$: $\mu_C = \mu_Y$ vs. $H_0$: $\mu_C > \mu_Y$

Conditions: We are told the treatments were randomly assigned to plots. Neither sample size is greater than 30. However, the boxplots do not suggest any strong skewness or outliers, so the *t*-procedure is appropriate.

Mechanics: $t = \dfrac{37.68 - 31.35}{\sqrt{\dfrac{4.44^2}{10} + \dfrac{7.74^2}{10}}} = 2.243$. With 14.35 degrees of freedom, the P-value is 0.021.

Conclusion: Since the p-value is less than a significance level of 5%, we have sufficient evidence to reject the null hypothesis. It appears CIF may be more effective in increasing tomato crop yield than YEC.

5. (a) the weight of the package will have mean μ=4(6)+5(8)+3=67 ounces and standard deviation $\sigma = \sqrt{4(.2)^2 + 5(.3)^2 + 1(.1)^2} = 0.79$ ounces.

(b) The package weights are N(67, 0.79). z = (68-67)/0.79 = 1.27. P(z>1.27) = 0.1020. There is a 10.2% chance a randomly selected package will require extra postage.

(c) The sampling distribution of the mean weights of a random sample of 4 boxes has mean 67 ounces and standard deviation 0.79/√4 = 0.395 ounces. z = (68-67)/0.395 = 2.53. P(z>2.53) = 0.0057. There is a 0.57% chance a random sample of four boxes will have an average weight greater than 68 ounces.

6. (a) Since there is a 10% chance of a no-show, let the digits 0-8 represent a person showing up for the flight and 9 represents a no-show. Starting with the first digit, read 12 digits and note the number of 9's. Continue reading blocks of 12 digits, recording the number of 9's each time until you have recorded 15 blocks. Count how many blocks had 0 or 1 no-shows. Divide this amount by 15 to determine the probability a flight will be overbooked.

(b)

1**9**223 **9**5034 05|756 28713 **9**640|**9** 12531 42544 8|2853 68417 350|13 1552**9** 47487
Trial 1: 2 noshow Trial 2: 1 noshow Trial 3: 1 noshow Trial 4: 0 noshow Trial 5: 1 noshow

73676 47150 **99**|400 01**9**27 2775|4 42648 82425 3|42**9**0 8273**9** 578|**9**0 20807 18883
Trial 6: 2 noshow Trial 7: 1 noshow Trial 8: 0 noshow Trial 9: 2 noshow Trial 10: 1 noshow

45467 7170**9** 77|558 000**95** 3286|3 2**9**485 82226 **9**|0056 60**9**40 720|24 17868 41**979**
Trial 11: 1 noshow Trial 12: 1 noshow Trial 13: 2 noshow Trial 14: 1 noshow Trial 15: 2 noshow

A total of 5 trials resulted in 2 no shows, while 7 trials resulted in 0 or 1. According to this simulation, the flight will be overbooked 7/12 = 58.33% of the time.

(c) If 90% of passengers showed up, we would expect 1.2 passengers to be no-shows, on average. This histogram suggests the average number of no-shows is closer to 3. It appears the actual percentage of passengers who show up may be lower than 90%.

## Practice Exam 2

### Section 1: Multiple-Choice

1. A child psychologist claims the mean attention span for a 1-year old is 12 seconds. An advocacy group claims exposure to a popular cartoon results in a decrease in attention span. To investigate this claim, which of the following hypotheses would be appropriate?

   (A) $H_0$: the mean attention span for children who watch the cartoon is less than 12 seconds.
   (B) $H_0$: the mean attention span for children who watch the cartoon is greater than 12 seconds.
   (C) $H_a$: the mean attention span for children who watch the cartoon is less than 12 seconds.
   (D) $H_a$: the mean attention span for children who watch the cartoon is greater than 12 seconds.
   (E) $H_a$: the mean attention span for children who watch the cartoon is less than or equal to 12 seconds.

2. In a random sample of 1500 adults taken in 1989, 551 adults indicated they were regular smokers in. In 1999, 2000 adults were randomly sampled and 652 indicated they were regular smokers. Which of the following is the 95% confidence interval for the difference in the true proportion of smokers between these two years?

   (A) $(0.367 \quad 0.326) \quad 1.960\sqrt{\frac{(0.367)(0.633)}{1500} \quad \frac{(0.326)(0.674)}{2000}}$

   (B) $(0.367 \quad 0.326) \quad 1.645\sqrt{\frac{(0.367)(0.633)}{1500} \quad \frac{(0.326)(0.674)}{2000}}$

   (C) $(0.367 \quad 0.326) \quad 1.960\sqrt{\frac{551}{2500} \quad \frac{652}{2500} \quad \frac{1}{1500} \quad \frac{1}{2000}}$

   (D) $(0.367 \quad 0.326) \quad 1.645\sqrt{\frac{551}{2500} \quad \frac{652}{2500} \quad \frac{1}{1500} \quad \frac{1}{2000}}$

   (E) $(0.367 \quad 0.326) \quad 1.960\sqrt{(0.367)(0.326) \quad \frac{1}{1500} \quad \frac{1}{2000}}$

3. Physical fitness of cardiac patients is measured by maximum oxygen uptake (milliliters per kilogram). The distribution of maximum oxygen uptake for cardiac patients is approximately Normal with mean 24.1 and standard deviation 6.3. How does the oxygen uptake for a patient in the $25^{th}$ percentile compare to that of an average patient?

   (A) approximately 0.67 ml/kg below the mean
   (B) approximately 5.63 ml/kg below the mean
   (C) approximately 5.63 ml/kg above the mean
   (D) approximately 4.22 ml/kg above the mean
   (E) approximately 4.22 ml/kg below the mean

4. Which of the following situations would be difficult to explore using a census?

   (A) You wish to know the proportion of teachers in a school district that have a Master's degree.
   (B) You want to know the average amount of time spent on homework by students in your high school.
   (C) You are interested in the difference in performance on the AP Statistics exam for students taught by two different teachers.
   (D) You want to know the proportion of homes in a suburban community that have wireless internet access.
   (E) You want to know the proportion of trees in a large state forest that are infected with Dutch Elm disease.

5. The following statistics provide a summary of the distribution of average SAT scores for the 50 states.

| Mean = 950 | Median = 930 | Standard Deviation = 30 | First Quartile = 895 | Third Quartile = 990 |
|---|---|---|---|---|

About 25 states have average SAT scores that are

(A) less than 895.
(B) less than 990.
(C) between 895 and 990.
(D) between 920 and 960.
(E) more than 950.

6. One week before an election, a news program asks viewers of its 5:00 PM broadcast to text their response ("Yes" or "No") to the following question, "Do you intend to vote for the incumbent candidate?" so the results can be shared on the 10:00 PM news. Which of the following is a reason for potential bias in this method?

(A) It should be done during both the morning and the 5:00 PM news broadcasts.
(B) People who feel strongly about the election are more likely to respond.
(C) It is being conducted too close to the election.
(D) The wording of the question is biased.
(E) There is no third option (like "Maybe") for viewers to text.

7. A random variable, $Y$, is Normally distributed with mean 500 and standard deviation 25. Which of the following is equivalent to the probability that $Y$ is less than 440?

(A) $P(Y > 440)$
(B) $P(Y \geq 440)$
(C) $P(Y > 560)$
(D) $P(440 < Y < 560)$
(E) $1 - P(Y < 440)$

8. A statistics teacher conducted an experiment with her class in which they used popsicle sticks and rubber bands to construct catapults for launching gummy bears. The students collected data on a large number of trials, noting how many rubber bands were used and how far the bear was launched (in feet). The relationship was found to be roughly linear, with a correlation coefficient of 0.921. What would happen to the correlation if the length measurements were converted to inches?

(A) Because there are 12 inches in a foot, the correlation coefficient will be 0.921/12 = 0.0768.
(B) Because there are 12 inches in a foot, the correlation coefficient will be 0.921 – 1/12 = 0.8377.
(C) Correlation is not affected by the units of measurement, so it will still be 0.921.
(D) You cannot change one measurement without changing the other. It is impossible to calculate.
(E) Because you can measure more accurately with inches, the correlation will close to 1.

9. The boxplots below show the reaction times (in milliseconds) for two groups of subjects who were exposed to either threatening or non-threatening stimuli.

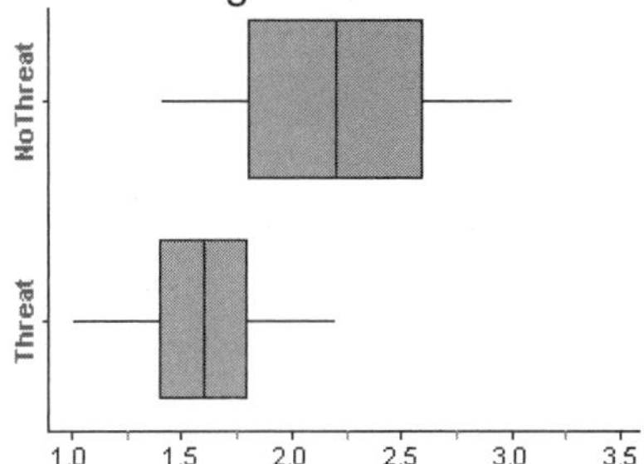


Which of the following statements is TRUE?

(A) The median reaction time of the group exposed to non-threatening stimuli is less than that of the group exposed to threatening stimuli.
(B) The range of the group exposed to non-threatening stimuli is less than that of the group exposed to threatening stimuli.
(C) The first quartile of the group exposed to non-threatening stimuli is greater than all reaction times for the group exposed to threatening stimuli.
(D) Half of the reaction times for the group exposed to non-threatening stimuli are greater than all of the reaction times for the group exposed to threatening stimuli.
(E) There were more subjects exposed to non-threatening stimuli than threatening stimuli.

10. A fair die is going to be rolled 6 times. The first 5 rolls are as follows: 3, 5, 1, 2, 4. What is the probability of rolling a 6 on the next roll?

(A) 1
(B) 1/6
(C) ${6 \atop 1}\frac{5}{6}^{5}\frac{1}{6}$
(D) $\frac{5}{6}^{5}\frac{1}{6}$
(E) 0

11. An agricultural researcher would like to investigate the effectiveness of a new pesticide in preventing insect damage to crops. To test the new pesticide against the current product in use, 12 plots are laid out in a field. Six of the plots are adjacent to a forest and 6 are not. What would be the best approach to assigning the treatments to the plots?

(A) Randomly assign the new pesticide to 6 of the 12 plots.
(B) Randomly assign the new pesticide to 3 of the plots adjacent to the forest and apply the current product to the other three. Repeat the random assignment with the six plots that are not next to the forest.
(C) Because insects are more likely next to the forest, assign the new pesticide to those 6 plots and the current product to the other 6.
(D) Assign both pesticides to each plot.
(E) You cannot carry out this experiment without a placebo.

12. The average ACT score for 250 randomly selected students planning on majoring in mathematics is 28.2 with standard deviation 1.3. The average ACT score for 200 randomly selected students planning on majoring in physics is 27.5 with standard deviation 1.6. A two-sided *t*-test results in the test statistic $t = 5.01$. What can you conclude from this test?

(A) The difference in means (0.7) is less than or equal to both standard deviations. Therefore there is not a significant difference between the mean ACT scores of the two types of students.
(B) We cannot conclude a difference exists because the groups did not have equal sample sizes.
(C) The physics students have a larger standard deviation. Therefore, they are more likely to have higher ACT scores.
(D) Because the *p*-value is less than 0.05, there is evidence of a significant difference between the mean ACT scores of the two types of students.
(E) Because the *p*-value is greater than 0.05, there is evidence of a significant difference between the mean ACT scores of the two types of students.

13. In which of the following distributions is the mean most likely greater than the median?

(A) 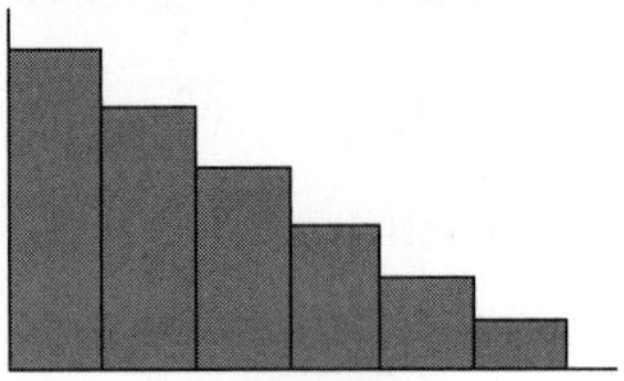

(B) 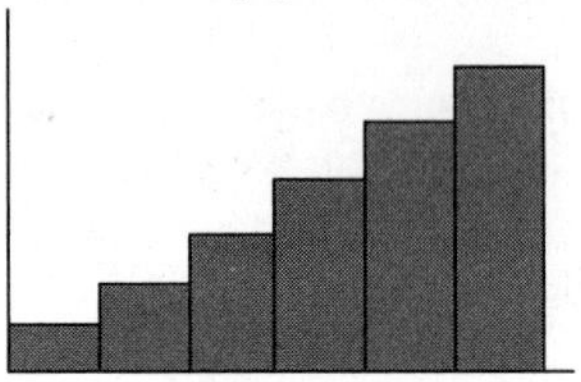

(C) 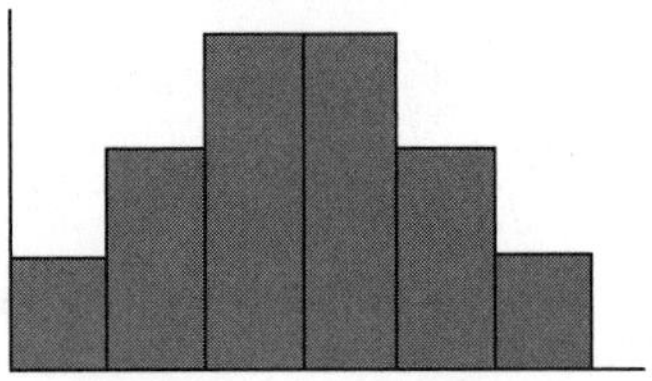

(D) 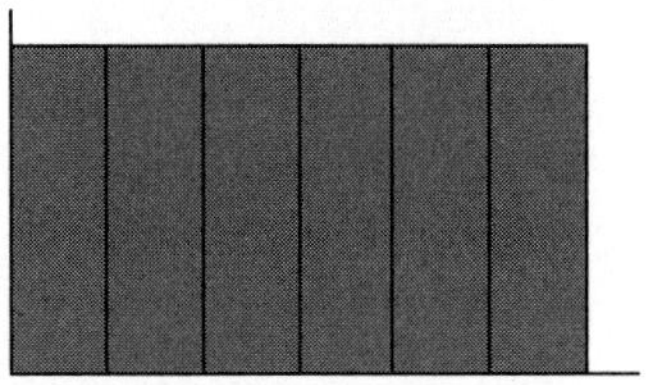

(E) 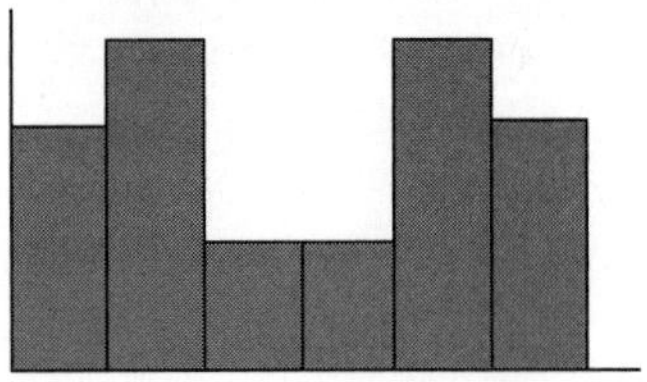

14. The weights of backpacks for students in two classes are measured. The backpacks of the 25 students in class A have an average weight of 12.3 pounds and a median weight of 14.2 pounds. The backpacks of the 28 students in class B have an average weight of 15.7 pounds and a median weight of 13.9 pounds. If the students were to combine classes, what would be the median weight of the 53 backpacks?

(A) $\dfrac{14.2 \quad 13.9}{2}$
(B) $\dfrac{12.3 \quad 15.7}{2}$
(C) $\dfrac{25(14.2) \quad 28(13.9)}{53}$
(D) $\dfrac{25(12.3) \quad 28(15.7)}{53}$
(E) The median cannot be calculated based on the information given.

15. A fair six-sided die is rolled 10 times and the number of 3's is recorded. This process is repeated 249 more times and the number of 3's is recorded each time. What kind of distribution has been simulated?

(A) Binomial: $n = 10$, $p = 0.3$
(B) Binomial: $n = 250$, $p = 1/6$
(C) Binomial: $n = 10$, $p = 1/6$
(D) The sampling distribution of sample proportions with $n = 10$ and $p = 0.3$
(E) The sampling distribution of sample proportions with $n = 10$ and $p = 1/6$

16. Which of the following is an advantage of an experimental study over an observational study?

(A) Experiments are easier to conduct since you control them.
(B) Fewer observations are needed to conduct an experiment.
(C) Because experiments involve comparing multiple groups, the conditions for inference are met more easily.
(D) Experiments can provide evidence of cause and effect relationships.
(E) Experiments take less time to conduct.

17. A study of black bears in northern Minnesota examined the relationship between a bear's neck girth (the distance around the neck in centimeters) and its weight (in kilograms). The correlation was calculated to be $r = 0.967$. Interpret the coefficient of determination for this relationship.

(A) There is a strong, positive, linear relationship between a bear's neck girth and its weight.
(B) Each increase of 1 centimeter in girth corresponds to an average increase of 0.967 pounds in weight.
(C) Approximately 93.5% of the variation in a bear's weight can be explained by the least squares regression on girth.
(D) Approximately 96.7% of the variation in a bear's weight can be explained by the least squares regression on girth.
(E) Each increase of 1 centimeter in girth corresponds to an average increase of 0.935 pounds in weight.

18. A study of two popular sleep-aids measured the mean number of hours slept for 200 individuals administered the drugs. In the group of 200 volunteer subjects, 100 were randomly chosen to be administered Drug 1 and the other 100 were administered Drug 2. A two-sample *t* test was performed on the difference in the mean sleep times experienced by the subjects. The *p*-value was 0.12. Which of the following is a correct interpretation of the *p*-value?

(A) Approximately 12% of the subjects did not sleep at all.
(B) There was a 12% increase in the mean sleep time by subjects in the study.
(C) There was a 12% difference between the mean sleep times of the two groups.
(D) Assuming both drugs are equally effective, we would expect to see a difference in the mean sleep times at least as extreme as the observed difference for 200 subjects 12% of the time.
(E) Approximately 12% of the subjects in the study experienced the same sleep time.

19. A random sample of a popular candy resulted in the following distribution of colors.

| Red | Green | Blue | Yellow |
|---|---|---|---|
| 40 | 25 | 25 | 20 |

A student performs a chi-square test on the hypotheses:
$H_0$: The proportion of candies is the same for each color.
$H_a$: At least one color's proportion differs from that of the other colors.
The test statistic is 8.182. What can this student conclude at a 5% level of significance?

(A) Reject the null hypothesis because 8.182 is greater than the critical value.
(B) Reject the null hypothesis because 8.182 is less than the critical value.
(C) Fail to reject the null hypothesis because 8.182 is greater than the critical value.
(D) Fail to reject the null hypothesis because 8.182 is less than the critical value.
(E) The test cannot be performed because not all colors have more than 30 observations.

20. A school principal is interested in calculating a 96% confidence interval for the true mean number of days students are absent during the school year. The attendance office lists the number of days absent for each student during the month of December. Which of the following is the best reason the principal cannot construct the confidence interval?

(A) A confidence interval cannot be constructed unless all data for the population are known.
(B) The critical value for 96% is not listed in the table, therefore it can't be calculated.
(C) The attendance office should report the average for the month, not the number for each individual student.
(D) Because most students wouldn't miss any days, the average will be too small to construct a confidence interval.
(E) The number of absences in December may not be representative of the rest of the months during the year. Data should be collected from all months.

21. When should a *t*-distribution be used for inference about a population mean?

(A) Whenever the sample size is small.
(B) Whenever a confidence interval for a mean is needed.
(C) Whenever the sample is small and the population standard deviation is known.
(D) Whenever the population is approximately Normal and the population standard deviation is estimated from a sample standard deviation.
(E) Whenever at least one condition for inference is violated.

22. The results of a recent test are displayed in the histogram below:

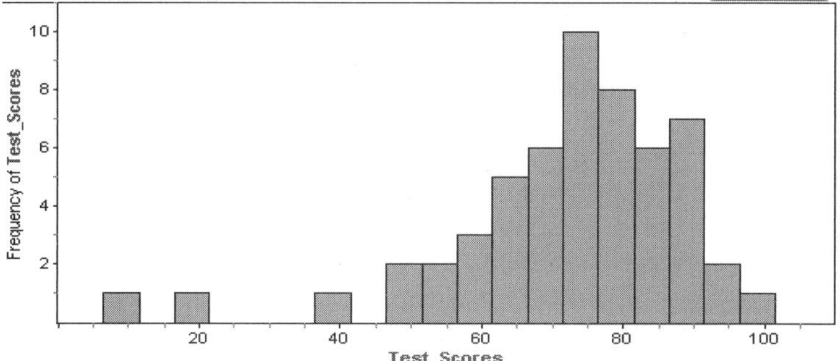


Which of the following represents a boxplot of the data?

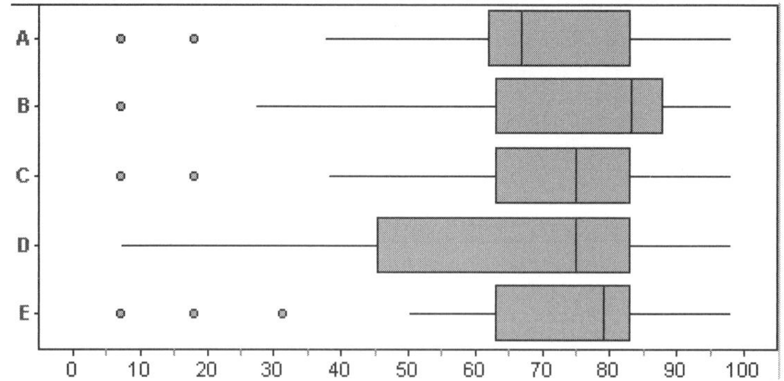

23. Because most chapter exams occur during the last week of the month, a teacher believes that students are more likely to be absent during that fourth week. The teacher collected data on the number of absences during each week of a random sampling of months over the past 5 years. The results are displayed below:

| Week | One | Two | Three | Four |
|---|---|---|---|---|
| Absences | 125 | 110 | 150 | 185 |

Which of the following would be the most appropriate inference procedure to test the teacher's claim?

(A) A chi-square test for association.
(B) A chi-square goodness-of-fit test.
(C) Multiple $z$-tests for proportions.
(D) A $t$-test for means.
(E) A linear regression $t$-test.

24. Scores on a standardized mathematics placement test are Normally distributed with mean 150 and standard deviation 10. Scores on a standardized science placement test are Normally distributed with mean 130 and standard deviation 15. Anne earned a score of 175 on the math test. If she had the same $z$-score on the science test, what was her score?

(A) 145
(B) 152.5
(C) 160
(D) 165
(E) 167.5

25. To determine the effectiveness of speed-boosting a computer by "overclocking" its processor, a study measured the difference between the mean speed (in MHz) for 10 computers that were "overclocked" and 10 identical computers that were not modified. The 95% confidence interval for the difference between the true mean speeds (not modified – "overclocked") was (1.3, 17.3). Based on this information, what can we conclude?

(A) The difference will be greater than 1.3 MHz approximately 95% of the time.
(B) The sample sizes are too small to draw a conclusion. We need more information.
(C) Since the interval does not contain 0, there is no significant difference between the speeds.
(D) 95% of the differences between speeds will be between 1.3 and 17.3 MHz.
(E) A two-sample *t* test should have been used instead of an interval.

26. Approximately 42% of teenagers admit they have texted while driving. A sample of 250 teens are randomly selected to take part in an online course designed to inform them of the dangers of distracted driving. Which of the following gives the mean and standard deviation of the sample proportion of teens who admit they have texted while driving?

(A) 105, 60.9
(B) 105, 7.80
(C) 0.42, 0.00097
(D) 0.42, 0.03121
(E) You cannot determine the mean and standard deviation from the information given.

27. A simple random sample of 50 adults is surveyed to determine the true proportion of adults who visit the dentist at least once per year and a confidence interval for the proportion is constructed. Suppose the researcher had surveyed a random sample of 450 adults instead and got the same sample proportion. How would the width of the confidence interval for 450 adults compare to that of the interval for 50 adults?

(A) The width would be about one-ninth the width of the original interval.
(B) The width would be about one-third the width of the original interval.
(C) The width would be the same as the width of the original interval.
(D) The width would be about three times the width of the original interval.
(E) The width would be about nine times the width of the original interval.

28. The number of taps per minute a person can do was measured before and after a person consumed a popular energy drink. The researcher is interested in the effect of energy drinks on a person's physical actions. The results are below:

| Subject | 1 | 2 | 3 | 4 | 5 | 6 | 7 | 8 | 9 | 10 |
|---|---|---|---|---|---|---|---|---|---|---|
| Before Drink | 105 | 99 | 93 | 96 | 95 | 99 | 94 | 87 | 100 | 89 |
| After Drink | 103 | 102 | 105 | 102 | 101 | 110 | 95 | 89 | 102 | 98 |

Which of the following tests would be most appropriate, assuming all conditions for inference are met?

(A) chi-square test for independence/association
(B) two sample *z*-test for proportions
(C) *t*-test for the slope of the regression line
(D) two-sample *t*-test for means
(E) matched pairs *t*-test for the mean difference

29. A two-proportion z-test of $H_0$: $p_1 - p_2 = 0$ vs. $H_a$: $p_1 - p_2 \neq 0$ results in a p-value of 0.052. Which of the following could be a 95% confidence interval for $p_1 - p_2$?

(A) (-0.48, -0.02)
(B) (-0.25, 0.01)
(C) (0.01, 0.25)
(D) (0.02, 0.34)
(E) We need to know the actual data.

30. Jason's golf scores are Normally distributed with mean 90 and standard deviation 6. Mark's golf scores are Normally distributed with mean 85 and standard deviation 8. Suppose Jason's and Mark's scores are independent. Which of the following would best describe the distribution between Mark's score and Jason's score (Mark's score – Jason's score)?

(A) Normal with mean 0 and standard deviation 1.
(B) Normal with mean -5 and standard deviation 2.
(C) Normal with mean -5 and standard deviation 10.
(D) Normal with mean -5 and standard deviation 14.
(E) Normal with mean 5 and standard deviation 14.

31. A popular fast-food restaurant chain wants to test-market a new sandwich option. The chain has approximately 3000 locations across the country. The restaurants are either downtown in urban cities, in shopping malls in suburban cities, or stand-alone restaurants in medium and small size cities. The chain randomly selects 10% of its downtown locations, 15% of its suburban locations, and 10% of its stand-alone locations in which to test the new sandwich option. What type of sampling method did the chain use?

(A) Simple random sampling
(B) Systematic sampling
(C) Convenience sampling
(D) Cluster sampling
(E) Stratified sampling

32. More and more adults are receiving their daily news in formats other than print and television. In a recent study, a random sample of 500 adults were asked to indicate their primary source of daily news: newspaper, television, radio, or online. A 95% confidence interval for the true proportion of adults who get their news online is given as (0.66, 0.74). Which of the following is TRUE?

(A) In repeated sampling of 500 adults, 95% of the intervals constructed will contain the true proportion of adults who get their news online.
(B) In repeated sampling of 500 adults, 95% of the sample proportions of adults who get their news online will fall in the interval (0.66, 0.74).
(C) In repeated sampling of 500 adults, the true proportion of adults who get their news online will fall in the interval (0.66, 0.74) 95% of the time.
(D) There is a 95% chance the true proportion of adults who get their news online falls in the interval.
(E) Approximately 95% of all samples of 500 adults will result in the interval (0.66, 0.74).

33. During WWII, German tanks were numbered 1, 2, 3, …, N. A group of captured tanks were considered as a random sample of tanks and their serial numbers used to estimate the number of tanks in the battlefield. Five estimators are constructed using different measures from the sample. The histograms below display the simulated sampling distributions for the five estimators. Which one is associated with the best estimator for N, the true number of German tanks? Assume the actual value of N is 342.

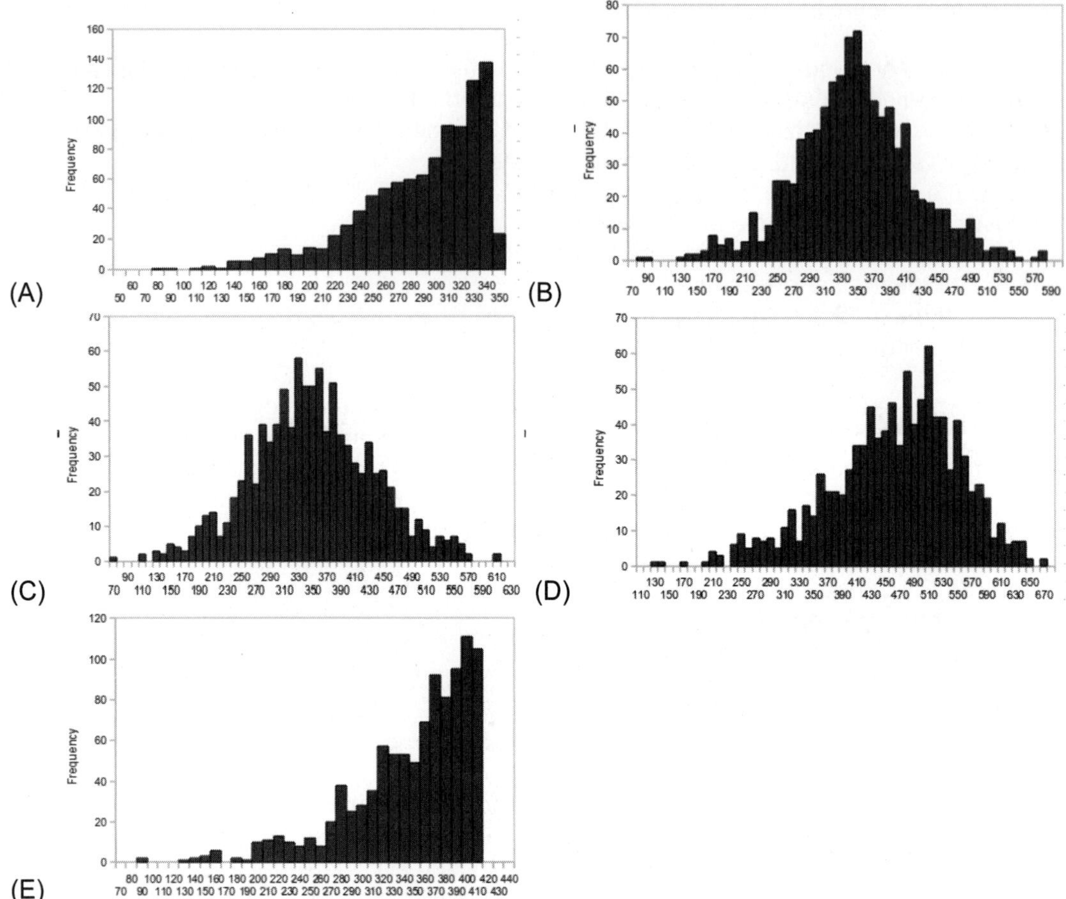


34. A researcher wants to measure the effectiveness of an exam preparation course. She will use a hypothesis test to detect differences between the true mean score of the population with a hypothesized mean score. If the difference is large, a small sample will suffice. If the difference is small, she will need to increase the sample size. This is an example of

(A) The power of a statistical test.
(B) A two-sample test statistic.
(C) A correlation coefficient.
(D) A Type II error.
(E) The *p*-value of a statistical test.

35. A drug company wants to test two new medications for high blood pressure. A sample of 200 volunteers with high blood pressure are to be randomly assigned a treatment for this study. Researchers believe the weight of the patients may play a role in the incidence of high blood pressure. Which of the following best describes how this experiment should be designed?

(A) Divide the volunteers into two groups: the 100 heaviest and the 100 lightest. Within each group, randomly assign one medication to 50 volunteers and the other 50 get the other medication.
(B) Divide the volunteers into two groups: the 100 heaviest and the 100 lightest. Flip a coin to determine which group gets which medication.
(C) Randomly assign one medication to 100 volunteers. The remaining 100 volunteers receive the other medication.
(D) Divide the volunteers into two groups: the 100 highest blood pressures and the 100 lowest. Flip a coin to determine which group gets which medication.
(E) Assign the first 100 volunteers who show up for the study to one medication and the last 100 to the other medication.

36. A recent story on the evening news cited a poll that said 4% of teachers in New Hampshire felt class sizes were too large while 36% of teachers in California felt the same way. The news anchor reported that an average of 20% of teachers in the two states feel class sizes are too large. Which of the following reasons explains why this conclusion is inaccurate?

(A) The polls did not represent an SRS of adults in each state.
(B) There is nothing wrong with this conclusion since (36% + 4%)/2 = 20%.
(C) The sample size from each state is not large enough to draw any conclusions.
(D) The number of teachers in each state is very different, therefore averaging is inappropriate.
(E) We cannot conclude anything since we don't know whether or not the sample sizes are equal.

37. A scatterplot is shown below. After calculating the least-squares regression line, it is determined that the point (13,15) was mistakenly entered. If this point is removed from the dataset, which of the following statements would be TRUE?

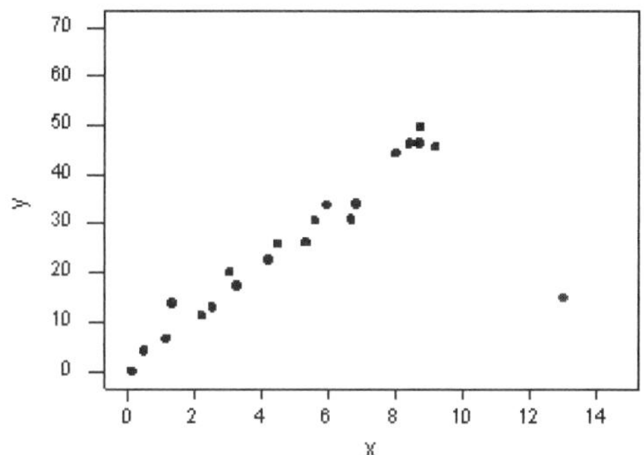


(A) The slope will increase and the correlation will decrease.
(B) The slope will increase and the correlation will increase.
(C) The slope will decrease and the correlation will remain the same.
(D) The slope will decrease and the correlation will decrease.
(E) The slope will decrease and the correlation will increase.

38. The probability distribution of $Y$, the number of practice problems assigned by a teacher each day, is given below

| $y$ | $P(Y=y)$ |
|---|---|
| 3 | 0.2 |
| 4 | 0.5 |
| 5 | 0.3 |

Which of the following represents the mean and variance, respectively, for $Y$?

(A) 4.00, 0.5
(B) 4.00, 1.25
(C) 4.10, 0.49
(D) 4.10, 0.70
(E) 4.10, 1.57

39. A researcher in Florida investigated the effects of age and gender on the short term memory of adults. In her study, the number of objects recalled in a memory test was recorded for a random sample of men (M) and a random sample of women (F).
The computer output below tests $H_0$: $\mu_M = \mu_F$ vs. $H_a$: $\mu_M \neq \mu_F$.

Two-sample T for M vs. F

| | N | Mean | StDev | SE Mean |
|---|---|---|---|---|
| M | 17 | 5.4 | 3.4 | 4.123 |
| F | 12 | 7.9 | 4.8 | 3.464 |

T=-1.550 P=0.1379

Assuming the conditions for inference have been met, which of the following conclusions can be drawn?

(A) The mean number of items recalled for men is significantly different than the mean number of items recalled for women at the 0.01 level.
(B) The mean number of items recalled for men is significantly different than the mean number of items recalled for women at the 0.05 level.
(C) The mean number of items recalled for men is significantly different than the mean number of items recalled for women at the 0.10 level.
(D) The mean number of items recalled for men is not significantly different than the mean number of items recalled for women at the 0.10 level.
(E) The mean number of items recalled for men is not significantly different than the mean number of items recalled for women at the 0.15 level.

40. Students at a suburban high school can enroll in several different math courses during their high school career. Suppose that 60% of all students enroll in AP Statistics, 12% enroll in AP Calculus, and 5% enroll in both. If a student is randomly selected, what is the probability he or she is not enrolled in either AP Statistics or AP Calculus?

(A) 0.05
(B) 0.28
(C) 0.33
(D) 0.48
(E) 0.60

## Practice Exam 2

### Section 2: Free-Response

1. Today's baseball players are bigger, faster, and stronger than baseball players from the early days of the game. However, there is some question as to whether or not they are better hitters. The following summary statistics show the number of home runs hit in a single season by 15 randomly selected All-Stars from the early 1900s and 12 randomly selected all stars from the 2000s.

| Time | Minimum | Q1 | Median | Q3 | Maximum | Mean | St Dev | N |
|---|---|---|---|---|---|---|---|---|
| 1900s | 22 | 35 | 46 | 54 | 60 | 43.93 | 11.25 | 15 |
| 2000s | 9 | 27 | 39 | 50.5 | 70 | 37.83 | 18.48 | 12 |

(a) Use this information to construct side-by-side boxplots that compare the number of home runs hit during each time period.

(b) Compare the distribution of home runs for each time period.

(c) If you wanted to determine if there was a significant difference in the number of home runs hit during the two time periods, what test of significance would you use? Explain your reasoning. Do not carry out the test.

(d) Given your choice in (c), are the conditions for the test met? Explain. Do not carry out the test.

2. A school district is interested in implementing a technology plan that would involve purchasing a tablet computer or a laptop computer for every student. The school board would like to determine the level of support for such a plan, so they decide to survey the public. The school board chairperson asks the first 250 community members who enter the school's homecoming game the following question

> Research indicates providing technology for students greatly increases their academic achievement. Do you think we should increase our students' achievement by providing tablet computers and laptops, even though it would result in a slight increase in your taxes?
> ____ Yes ____ No ____ No opinion

(a) Explain why this sampling method may introduce bias into the results. Do you think the sample would over- or under-estimate the actual proportion of the community in favor of the plan? How should the school board have selected their sample to avoid this bias?

(b) Explain why this question may be biased. How could it be worded to avoid that bias?

3. A simple random sample of 592 students was selected. The eye color and natural hair color for each student was recorded. The resulting data are summarized in the following table.

| | Hair Color | | | | |
|---|---|---|---|---|---|
| Eye Color | Black | Brown | Red | Blonde | Total |
| Brown | 68 | 119 | 26 | 7 | 220 |
| Blue | 20 | 84 | 17 | 94 | 215 |
| Hazel | 15 | 54 | 14 | 10 | 93 |
| Green | 5 | 29 | 14 | 16 | 64 |
| Total | 108 | 286 | 71 | 127 | 592 |

(a) What is the probability that a person chosen at random from this sample will have brown hair?

(b) What is the probability that a person chosen at random from this sample who has blue eyes will also have brown hair?

(c) Based on your answers to (a) and (b), are hair color and eye color independent?

4. The state of the economy is an important issue during election years. A political analyst believes half of all consumers are optimistic the economy is improving. To test this, 484 consumers were randomly selected and asked their opinion of the state of the economy. From the sample, 257 reported they were optimistic that the economy was improving.

(a) Do these data provide evidence that the proportion of consumers who feel the economy is improving is greater than 0.5? Conduct an appropriate test of significance to justify your answer.

(b) Interpret the p-value from your test of significance in the context of the problem.

5. An automated manufacturing process involves four independent steps. The total production time is the sum of the 4 individual times for each step of the process. The times to complete each step (in minutes) are normally distributed with the following means and standard deviations.

| | Mean | St Dev |
|---|---|---|
| Step 1 | 7.1 | 0.24 |
| Step 2 | 6.8 | 0.25 |
| Step 3 | 6.6 | 0.23 |
| Step 4 | 7.0 | 0.24 |

(a) What is the probability that Step 3 will take less than 6.3 minutes to finish?

(b) What are the mean and standard deviation of the total time necessary to complete the process?

(c) The process is monitored occasionally to determine how fast it is taking to complete. If the process is observed to take more than 29 minutes, it is shut down and the individual steps are adjusted. What is the probability that a random observation will result in an adjustment?

6. An industrial psychologist is interested in studying the effect of incentives on worker productivity. In his study, a sample of workers at two manufacturing plants were randomly assigned to earn bonus incentives of 20¢, 30¢, or 40¢ for each item produced above their weekly quota. The productivity of the workers was measured in the number of items produced each week.

The prediction model for productivity (y) based on incentive (in cents per item above weekly quota) is given below.

Model 1: $\hat{y}\quad 1389.72\quad 6.23(\textit{incentive})$

(a) Interpret the slope of this model in context.

(b) Use Model 1 to predict the productivity when an incentive of 25¢ is offered.

After studying the results, the researcher notices half of the workers were part of a union and half were non-union workers. To explore whether or not there is a difference in the productivity for the two types of workers when an incentive is added, he constructs a new model with an indicator variable, named *type*. This variable takes a value of 1 when the worker is a non-union member and 0 when the worker is a union member. The newly created prediction model is given below.

Model 2: $\hat{y}\quad 1365.83\quad 6.217(\textit{incentive})\quad 47.778(\textit{type})\quad 0.033(\textit{incentive})(\textit{type})$

(c) Use Model 2 to determine a prediction model for union workers and a prediction model for non-union workers. Sketch these models to illustrate and comment on the difference.

(d) Use Model 2 to predict the productivity for each type of worker when an incentive of 25¢ is offered.

**Answer Key**
**Practice Exam 2**
**Multiple-Choice**

1. The null hypothesis is a statement of equality. The alternative hypothesis is a statement of the claim for which you are gathering evidence. Ans: C

2. The confidence interval for the difference between two proportions is given by $\hat{p}_1 \quad \hat{p}_2 \quad z^* \sqrt{\frac{\hat{p}_1(1 \quad \hat{p}_1)}{n_1} \quad \frac{\hat{p}_2(1 \quad \hat{p}_2)}{n_2}}$. For a 95% confidence interval, z* = 1.960. Ans: A

3. The $25^{th}$ percentile corresponds to a z-score of approximately -0.67. A patient in the $25^{th}$ percentile would be 0.67 standard deviations below the mean. 6.3(0.67)=4.22 Ans: E

4. A census involves measuring every element in the population. It would be difficult to check every tree in a large state forest. Ans: E

5. Since half of the states will fall within the interquartile range, 25 states will have average scores between the first and third quartiles. Ans: C

6. This is a voluntary response survey, which favors those who feel more strongly about the issue. Ans: B

7. The value 440 is 60 below the mean. Since the distribution is symmetric, the probability *Y* is less than 440 will be equivalent to the probability *Y* is greater than 500+ 60 = 560. Ans: C

8. Correlation is not affected by units of measurement, so it will remain the same. Ans: C

9. The median of the non-threatening group is equal to the maximum of the threatening group. All reaction times for the threatening group are less than the median of the non-threatening group. Ans: D

10. Each roll of a die is independent of previous rolls, so the probability of rolling a particular number does not change dependent upon what has been rolled. Ans: B

11. Blocking would be appropriate since the plots next to the forest may be exposed to more insects, etc. Both treatments should be randomly assigned to the plots next to the forest. Ans: B

12. The *p*-value for this test is $8.5 \times 10^{-7}$. Because this is less than any reasonable significance level, there is evidence of a significant difference between the means. Ans: D

13. The mean is greater than the median in a skewed right distribution. Ans: A

14. Without knowing the actual values, the median cannot be calculated. Ans: E

15. We are interested in the number of successes out of 10 rolls. This is a binomial experiment with $n$ = 10 and probability of success = 1/6. Ans: C

16. Cause and effect relationships can only be established by experiments. Ans: D

17. The coefficient of determination is $r^2 = 0.935$. This value represents the proportion of variation in predicted *y*-values that can be explained by the least squares regression on *x*. Ans: C

18. The *p*-value tells us how likely it would be to observe a result at least as extreme as the observed result, assuming the null hypothesis is true. Ans: D

19. The test statistic is greater than the critical value for 5% level of significance. Therefore the *p*-value is less than 5%. Reject the null hypothesis. Ans: A

20. Absences during December may not be representative of absences during the entire school year. Ans: E

21. The *t*-distribution is used when the population is assumed to be approximately Normal and the population standard deviation is unknown. Ans: D

22. The median appears to be about 75 and there are two potential outliers. Ans: C

23. We have one categorical variable of interest. A goodness-of-fit test is appropriate. Ans: B

24. Anne's math score is z = (175 – 150)/10 = 2.5. Therefore, Anne's science score is 130 + 2.5(1.5) = 167.5 Ans: E

25. The sample sizes are less than 30. Without knowing the actual data, we cannot draw a conclusion from the interval. Ans: B

26. This is a binomial distribution with $n$ = 250 and $p$ = 0.42. The sampling distribution of the sample proportion has a mean $\mu$ = 0.42 and standard deviation

$$\sqrt{\frac{p(1\quad p)}{n}}\quad\sqrt{\frac{.42(.58)}{250}}\quad 0.0312$$

Ans: D

27. The margin of error of a confidence interval for a proportion is $ME\quad z\quad\sqrt{\frac{\hat{p}(1\quad\hat{p})}{n}}$. If the sample size is nine times the original, the margin of error will increase by a factor of 3. Ans: B

28. Since each subject was administered each "treatment", a matched pairs test would be most appropriate. Ans: E

29. Since the test is not significant at the 5% level of significance, a 95% confidence interval would not contain 0. Ans: B

30. Since both distributions are Normal, the resulting distribution of *M-J* will be Normal with mean $\mu_M - \mu_J$. The standard deviation is found by taking the square root of the sum of the variances. $\sqrt{6^2\quad 8^2}\quad 10$. Ans: C

31. In stratified sampling, the population is broken down into "strata" and random samples from each stratum are selected. Ans: E

32. The confidence level tells us how confident we are in the method used. That is, it tells us the "success rate" of the method in repeated sampling. Ans: A

33. The best estimator will be unbiased and have low variability. The mean of the sampling distribution should be around 342. The distributions in B, C, and E all have means close to 342. The distribution in E would have a lower standard deviation than those in B and C. . Ans: E

34. Increasing sample size increases the "power" of a test to detect differences. Ans: A

35. Since weight may affect blood pressure, the experiment should be blocked by weight. Ans: A

36. New Hampshire has a much smaller population than California and, therefore, fewer teachers. Since the populations are so different, averaging is inappropriate. Ans: D

37. The point is influential and will "pull" the line towards it. Removing the point will cause both the slope and the correlation to increase. Ans: B

38. $E(Y) = 0.2(3)+0.5(4)+0.3(5) = 4.10$
$VAR(Y) = (0.2)(3-4.1)^2+(0.5)(4-4.1)^2+(0.3)(5-4.1)^2=0.49$ Ans: C

39. Since the p-value is 0.1379, there is significant evidence of a difference at the 0.15 level, but not the 0.10 level. Ans: D

40. The probability is 1 – P(S or C) = 1 – { P(S) + P(C) – P(S and C) } = 1 – (0.60 + 0.12 – 0.05) = 0.33 `Ans: C

**Answer Key**
**Practice Exam 2**
**Free Response**

1. (a)

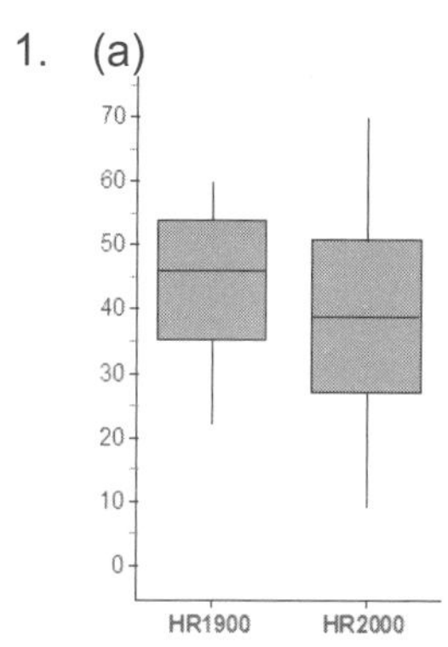


(b) The distribution of home runs hit in the 1900s is slightly skewed left, while the distribution for the 2000s is symmetric. The median number of home runs hit in the 1900s is higher than that of the 2000s, while the maximum number of home runs hit in the 2000s is much higher than that of the 1900s. The distribution of home runs hit in the 2000s has a larger range and standard deviation than the distribution for the 1900s. It is difficult to say if one era had a higher number in general since there is a great deal of overlap between the two distributions.

(c) Since we are comparing counts of home runs, we should compare the average number of home runs hit during each era. This could be accomplished by conducting a two-sample *t*-test for comparing means.

(d) The conditions are met. We are told the data are from two random samples of All-Stars from the early 1900s and early 2000s. We can assume these samples are independent. Both sample sizes are less than 30. However, the boxplots do not indicate strong skewness or outliers in either distribution, so we are safe to use the *t*-procedures.

2. (a) This is a convenience sample, so it may not be representative of the entire community. Since individuals attending the homecoming game may be more involved in or supportive of the school, they may be more likely to support a plan such as this. The sample would most likely overestimate the support for the plan. Instead, the school board should randomly select voters in the community to participate in the survey.

(b) This question suffers from wording bias. The first statement makes a person more likely to select *yes* since few people would be against increasing student achievement. Further, the word "*slight*" downplays the fact that by passing this plan, residents would have to pay more in taxes. A better wording would simply be "Do you support a technology plan that would provide all students with a tablet computer or a laptop?"

3. (a) P(brown hair) = 286/592 = 0.4831

(b) P(brown hair | blue eyes) = 84/215 = 0.3907
(c) If hair color and eye color are independent, then the two probabilities in (a) and (b) should be approximately the same. That is, P(brown hair) should equal P(brown hair | blue eyes). However, the two probabilities are not equal. Therefore hair color and eye color are not independent.

4. (a) Let $p$ = the true proportion of consumers who are optimistic the economy is improving. We will use a $z$-test for a proportion to test $H_0$: $p$=0.5 vs. $H_a$: $p$>0.5.

Conditions: We are told this is a random sample. The conditions for this test are $np_0 \geq 10$ and $n(1-p_0) \geq 10$. Both $np_0 = 242$ and $n(1-p_0) = 242$ are greater than 10. There are more than 4840 consumers.

Mechanics: $z \quad \frac{(0.53 \quad 0.50)}{\sqrt{\frac{0.5(0.5)}{484}}} \quad 132$ and P-value = 0.0934.

Conclusion: Since the p-value is greater than a significance level of 5%, we do not have sufficient evidence to reject the null hypothesis. We can not conclude that the proportion of consumers who are optimistic the economy is improving is greater than 0.50.

(b) The p-value is 0.0934. This means that, assuming the actual proportion of consumers who feel the economy is improving is 0.50, 9.34% of all random samples of 484 consumers will result in a proportion at least as extreme as the observed proportion.

5. (a) z = (6.3-6.6)/0.23 = -1.30. P(z<-1.30) = 0.0968. There is a 9.68% chance that Step 3 will be completed in less than 6.3 minutes.

(b) mean = 7.1+6.8+6.6+7 = 27.5 minutes. Std dev = $\sqrt{0.24^2 \quad 0.25^2 \quad 0.23^2 \quad 0.24^2} \quad 0.48$ minutes.

(c) z = (29-27.5)/0.48 = 3.125. P(z>3.125) = 0.0009. There is a 0.09% chance that the procedure will need to be adjusted based on a random observation of its completion time.

6. (a) The slope is 6.23. This means we predict an increase in productivity of about 6.23 units per week for each additional cent of incentive offered.

(b) The predicted productivity will be 1389.72+6.23(25) = 1545.47 units.

(c) Union workers (type = 0):
productivity = 1365.83+6.217(incentive)
Nonunion workers (type = 1):
productivity = 1365.83 + 6.217(incentive) + 47.778(1) + 0.033(incentive)(1)
productivity = 1413.61 + 6.25(incentive)

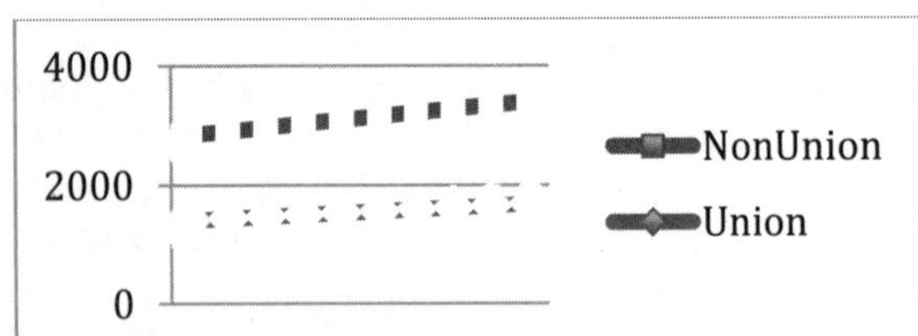


It appears the incentive has a greater effect on nonunion workers. Their productivity is higher and the greater slope suggests we predict a bigger increase in productivity for each additional cent in incentives.

(d) Union productivity = 1365.83 + 6.217(25) = 1521.255 units
Nonunion productivity = 1413.61 + 6.25(25) = 1569.86 units